ACTIVATION OF SATURATED HYDROCARBONS BY TRANSITION METAL COMPLEXES

CATALYSIS BY METAL COMPLEXES

A. E. SHILOV

The Institute of Chemical Physics of the USSR Academy of Sciences

ACTIVATION OF SATURATED HYDROCARBONS BY TRANSITION METAL COMPLEXES

D. REIDEL PUBLISHING COMPANY

A MEMBER OF THE KLUWER  ACADEMIC PUBLISHERS GROUP

DORDRECHT / BOSTON / LANCASTER

CHEMISTRY

Library of Congress Cataloging in Publication Data

CIP

Shilov, A. E. (Aleksander Evgenievich), 1930–
    Activation of saturated hydrocarbons by transition metal complexes.

    (Catalysis by metal complexes)
    Bibliography: p.
    Includes index.
    1. Catalysis. 2. Hydrocarbons. 3. Transistion metal compounds.
4. Complex compounds. I. Title. II. Series.
QD505.S53   1984   547′.010459   84–6817
ISBN 90-277-1628-5

Published by D. Reidel Publishing Company,
P.O. Box 17, 3300 AA Dordrecht, Holland.

Sold and distributed in the U.S.A. and Canada
by Kluwer Academic Publishers
190 Old Derby Street, Hingham, MA 02043, U.S.A.

In all other countries, sold and distributed
by Kluwer Academic Publishers Group,
P.O. Box 322, 3300 AH Dordrecht, Holland.

Printed in the Netherlands

CONTENTS

PREFACE   ix
    List of Abbreviations   x

INTRODUCTION   1
    References   10

CHAPTER I.   *Reactions of Metal Complexes with Compounds Containing 'Activated' C—H Bonds*   11
I.1.   Aromatic Hydrocarbons   11
    I.1.1.   Reactions of Arenes with Electrophilic Oxidants   11
    I.1.2.   Mercuration and Similar Reactions   12
    I.1.3.   Oxidation of Aromatic Molecules by Bivalent Palladium   14
    I.1.4.   Reactions of Arenes in Platinum Salt Solutions   17
    I.1.5.   Reactions of Arenes with Low Oxidation State Metal Complexes   17
    I.1.6.   Isotope Exchange of Arenes with Deuterium   19
I.2.   Reactions of 'Activated' Aliphatic C—H Bonds   20
    I.2.1.   C—H Bonds in the α-Position of the Side Chain of the Aromatic Ring: Reactions with Oxidants   21
    I.2.2.   Reactions of Alkylaromatic Compounds in the Presence of Platinum(II) Complexes   23
    I.2.3.   Reactions of Toluene in the Presence of Palladium Complexes   23
    I.2.4.   Activation of C—H Bonds by Polar Substituents   24
    I.2.5.   Activation of C—H Bonds in Ligands of Metal Complexes   26
References   32

CHAPTER II.   *Reactions of Alkanes with Compounds Other Than Metal Complexes*   35
II.1.   Reactions of Alkanes with Electrophiles   35
II.2.   Reactions of Alkanes with Atoms and Free Radicals   40

II.3.   Reactions between Alkanes and Carbenes   43
II.4.   Alkane Hydroxylation by Peracids   45
References   47

CHAPTER III. *Activation of Alkanes on the Surface of Metals and*
              Metal Oxides. Reactions of Alkanes with Metal Atoms
              and Ions   49
III.1.   Isotope Exchange   50
III.2.   Dehydrogenation and Dehydrocyclization   55
III.3.   Heterogeneous Oxidation   57
III.4.   Reactions of Alkanes with Free Metal Atoms and Ions   58
References   60

CHAPTER IV. *Homogeneous Oxidation of Alkanes in the Presence of*
              Metal Compounds   62
IV.1.   Direct Oxidation of Alkanes by Metal Compounds and Com-
         plexes in High Oxidation State   62
         IV.1.1.   Alkane Oxidation by Compounds of Chromium(VI)
                   and Manganese(VII)   63
         IV.1.2.   Complexes of Ruthenium(IV) and Iridium(IV)   65
         IV.1.3.   Peculiarities of Alkane Oxidation by Compounds of
                   Cobalt(III). Other Oxidants   67
         IV.1.4.   Oxidation of Alkanes in Concentrated Solutions of
                   Sulfuric Acid   70
         IV.1.5.   Palladium(II) Complexes   73
IV.2.   Participation of Transition Metal Ions and Complexes in the
         Oxidation of Hydrocarbons by Molecular Oxygen   74
         IV.2.1.   Chain Mechanism of Oxidation   74
         IV.2.2.   Catalytic Oxidation of Hydrocarbons in the Presence of
                   Small Concentrations of Transition Metal Ions   77
         IV.2.3.   Catalytic Oxidation of Hydrocarbons in the Presence of
                   Large Concentrations of Transition Metal Compounds.
                   Catalyst Participation in Chain Propagation   81
IV.3.   Biological Oxidation of Alkanes   88
         IV.3.1.   α-Ketoglutarate-Coupled Dioxygenases   90
         IV.3.2.   Cytochrome P-450   92
         IV.3.3.   Methanemonooxygenase   99
         IV.3.4.   On the Mechanism of Biological Oxidation   101
IV.4.   Coupled Oxidation of Hydrocarbons and Metal Complexes.
         Chemical Models of Monooxygenases   103

IV.4.1.   Hydroxylation of Hydrocarbons Coupled with Oxida-
tion of Metal Ions and Complexes   106
IV.4.2.   The Mechanism of Stannous Chloride Autoxidation   113
IV.4.3.   The Mechanism of Cyclohexane Oxidation Coupled
with Oxidation of $SnCl_2$ 116
IV.4.4.   Complexes of Molecular Oxygen with Transition Metal
Compounds and their Reactivity   119
IV.4.5.   Peroxides and Iodosobenzene as Oxidants: Models of
Oxygenase Active Center   121
IV.5.   On the Mechanism of Alkane Reactions with Electrophilic
Oxidants. General Considerations   125
References   137

CHAPTER V.   *Activation of Alkanes by Metal Complexes of Medium
and Low Oxidation State*   142
V.1.   General Remarks   142
V.2.   Reactions of Arenes and Alkanes in the Presence of Platinum(II)
Complexes   145
V.2.1.   H–D Exchange with the Solvent   145
V.2.2.   Kinetics of H–D Exchange   147
V.2.3.   Relative Reactivity of Hydrocarbons in H–D Exchange   151
V.2.4.   Multiple Exchange and its Mechanism   156
V.2.5.   Oxidation and Dehydrogenation of Alkanes in the
Presence of Pt(II) and Pt(IV) Complexes: Products   163
V.2.6.   Reactivity of Different Hydrocarbons. Mechanism of
Oxidation   165
V.2.7.   The Isolation of Aryl and Alkyl Platinum Complexes
in Reactions of Platinum Halides with Hydrocarbons
and their Role as Intermediates   172
V.2.8.   General Considerations of the Reactions of Hydro-
carbons with Pt(II) Complexes   178
V.3.   The Interaction of Alkanes with Complexes of Other Platinum-
Group Metals   182
V.3.1.   H–D Exchange   182
V.3.2.   The Dehydrogenation of Alkanes by Ir(I) Complexes   183
V.3.3.   Oxidative Addition of C–H Bonds in Alkanes to
Photochemically Generated Iridium(I) Complexes   185
V.4.   The Activation of Alkanes by Low-Valent Complexes of Non-
Platinum Metals   185
V.4.1.   Ziegler-Natta Systems   186
V.4.2.   Addition of Methane to Unsaturated Molecules   187

V.4.3. Dehydrogenation of Alkanes in the Presence of
        Rhenium Hydride Complex   189
V.5. A Comparison of Homogeneous and Heterogeneous Activation
      of Hydrocarbons   190
References   193

CONCLUSION   195

INDEX   197

PREFACE

The studies in coordination catalysis in homogeneous solutions in the last few decades have led to the discoveries of catalytic reactions of various types of molecules, including dihydrogen, acetylenes, olefins, aromatics, carbon monoxide, carbon dioxide, dioxygen, dinitrogen and other molecules. In such reactions the substrate molecule, or the fragment of it which enters the coordination sphere of metal complexes as a ligand, is chemically activated, i.e. it acquires the ability to become involved in reactions which do not occur at all, or proceed very slowly, without the metal complex.

Recently, alkanes have been added to the list of molecules which can be 'activated' by metal complexes, although many reactions of alkanes have been known for decades (for example, radical-chain reactions involving chlorination or oxidation, reactions with superacids, surface reactions on metal or metal oxides, etc.).

The reactions of alkanes with metal compounds were earlier restricted to the initiation of radical-chain oxidation and to interactions with such strong oxidants as $Mn(VII)$, $Cr(VI)$ or $Co(III)$, where the initial formation of free radicals leads further to subsequent, rather non-selective, reactions.

The great interest in catalytic activation of alkanes is explained primarily by the need to find new selective industrial ways to process hydrocarbons.

This apart, the saturated character coupled with a certain chemical inertness of alkanes create serious difficulties in their activation by metal complexes, this presenting an interesting theoretical problem.

Since the discovery of the first reactions of metal complexes with alkanes, several groups of workers have achieved important results, and this has allowed the activation of alkanes by metal complexes to form a new field in coordination catalysis. Although the alkane reactions which can be catalyzed by metal complexes are not so numerous as, for example, dihydrogen reactions, their study is promising and the field is developing successfully. A new decade may lead to interesting new achievements. The author would like to hope that this monograph will contribute to the development of this field which becomes more and more significant as we realize the utmost necessity of economic consumption of hydrocarbons in coal, petroleum and natural gas which have been stored by

ix

nature for millions years of the earth's existence, and are now so carelessly being consumed by man.

I would like to thank N. G. Vorobjeva for technical assistance in preparing the book for publication and L. I. Karkovsky for drawing some of the figures.

I am extremely grateful to the colleagues who have discussed with me the problems which are covered in this book, and who have helped to remove some of the errors and shortcomings; in particular, I thank Dr. Yu. V. Geletii who helped me with the problems of oxidation in the liquid phase, Dr. A. M. Khenkin who had made a valuable contribution to my knowledge of chemical models of biological oxidation, Prof. A. I. Archakov, Prof. G. I. Likhtenstein, and Dr. R. I. Gvozdev, who discussed the mechanisms of biological oxidation with me, Dr. A. A. Shteinman who for a long time was my colleague in the work on the activation of alkanes by platinum complexes, Dr. A. F. Shestakov who helped me to understand the theoretical side of the problem.

I am also grateful to many of my distinguished friends and colleagues in the field of coordination chemistry, particularly Prof. J. Halpern, Dr. G. W. Parshall, Prof. J. T. Groves, Prof. I. I. Moiseev, Prof. E. S. Rudakov, and many others who discussed with me the problems of activation of alkanes and helped to correct and develop my views in this rather complicated field.

Naturally, none of them is responsible for the shortcomings in the book which have no doubt remained. I hope to be able to correct at least some of them in future.

INTRODUCTION

The well-known chemical inertness of alkanes is reflected in their old names: paraffins (from Latin: *parum affinis* – devoid of affinity) and saturated hydrocarbons, i.e. hydrocarbons incapable of binding.

Reactions of alkanes taking place in mild conditions, not involving transition metals, usually proceed readily with short-lived species which possess high reactivity, such as free atoms and radicals, carbenes and their analogs

$$R \cdot + R'H \longrightarrow RH + R'$$

$$RH + CH_2 \longrightarrow RCH_3$$

$$RH + O \longrightarrow ROH$$

As compared with other, more chemically active molecules, alkanes have no multiple bonds or lone electron pairs, while their non-polar σ-C—H and σ-C—C bonds are usually very stable.

Table I lists some characteristics of alkanes and, for comparison, of other molecules. When reviewing the data given in Table I, it becomes clear why the alkanes are chemically inert. Their C—H bond energies (D(C—H)) as well as ionization potentials (IP) are high, electron affinities (EA) are negative values, proton affinities (PA), though positive, are far lower than for such hydrocarbon molecules as ethylene and benzene. Alkane acidities are many orders of magnitude smaller that those of acetylene and dihydrogen, let alone such molecules as alcohol and water.*

However, according to the data listed in Table I, dihydrogen, judging by some of its parameters, should react with even more difficulty than the alkanes, although the activation of H_2 on metal surfaces and in solutions of metal complexes (e.g., in the case of hydrogenation) is well known to proceed so readily as to allow the process often to occur at room and even lower temperatures. Although the C—H bond energy in benzene and ethylene is higher than in alkanes, these bonds are broken in a number of catalytic reactions with the

---

* According to different estimates C—H acidities of hydrocarbons differ significantly and real acidities could be even lower than those in Table I [1]. However, the order of acidities for various hydrocarbons apparently follows that of the Table.

1

TABLE I
Some characteristics of alkanes and other molecules

RH molecules	R	$D(R-H)$ kcal/mole	IP(eV)	EA(eV)	PA(eV)	pK_a
H_2	H	104	15.4	−3.6	4.4	25
CH_4	CH_3	104	12.7		5.3	40
C_2H_6	C_2H_5	98	11.5		5.6	42
C_3H_8	n-C_3H_7	97	11.1		6.1	
	iso-C_3H_7	94				44
C_4H_{10}	$tert$-C_4H_9		10.6			
C_6H_{12}	C_6H_{11}	94	9.9			45
C_6H_6	C_6H_5	109	9.2	−1.10	7.5	37
C_2H_2	C_2H	120	11.4			25
C_2H_4	C_2H_3	106	10.5	−1.81	6.9	36.5
CH_3OH	CH_3O	92	10.9		7.8	16
CH_3CN	CH_2CN	79	12.1		8.1	24
$CH_3CH=CH_2$	$CH_2-CH=CH_2$	86	9.7		7.9	35.5
$C_6H_5CH_3$	$C_6H_5CH_2$	85	8.8	−1.3	7.3	35
H_2O	OH	118	12.6	1.8	7.14	15.7

participation of metal complexes. Hence, it is interesting to compare the data on alkanes and molecular hydrogen as well as other molecules and to consider possible kinds of mechanisms of reactions with a variety of metal complexes.

In the case of alkanes the interaction is in fact limited to the cleavage of covalent σ-C—H or σ-C—C bonds. Moreover, taking into consideration the effect of steric hindrance in the reactions of metal complexes, the interaction with the C—H bond seems to be preferred and it is, naturally, the only possible interaction for the first alkane in the series, i.e. methane.

The relative reactivity of alkanes and dihydrogen as well as the selectivity of the C—H bond attack towards nucleophilic and electrophilic reactions (for example, H^+ and H^- elimination) must change in the opposite order:

$$E \quad H_2 < CH_4 < C_2H_6 < C_3H_8 \ldots 1° < 2° < 3°$$
$$N \quad H_2 > CH_4 > C_2H_6 > C_3H_8 \ldots 1° > 2° > 3°$$

($1°$, $2°$, $3°$ are primary, secondary and tertiary C—H bonds, respectively). These orders correspond to the opposite tendencies in ionization potentials, proton affinities and kinetic acidities (see Table I).

It should be taken into consideration that the most interesting case would be the reaction without the strong inhibition produced by the interaction of the alkane with the catalyst, by the other reagents present, and by the possible products formed due to their own competitive interaction with the metal complex. In this connection, it is desirable that the complex activating the alkane should not be too strong an oxidant (or reductant) and not too strong an acid or base. Besides this, we would naturally like the complex to have true catalytic properties, drawing the alkane into a particular chemical reaction with catalyst regeneration.

For example, if we consider some catalytic oxidation, then the oxidation of the alkane by a catalyst should be followed by catalyst regeneration, i.e. in this case by a reoxidation of the catalyst by a reagent. This is evidently not very easy to achieve if the catalyst itself is an exceptionally strong oxidant.

Another problem is the choice of solvent which, if it too posseses C—H bonds, can very naturally react with the catalyst present in the solution, in competition with the alkanes.

In their reactions with alkanes all the metal complexes may be divided into those of high, medium* and low oxidation states.

Metal complexes in a high oxidation state can interact with alkanes as oxidants or as hard acids, and the increase of their acidic properties must increase their oxidative properties. The high ionization potential of the alkanes must be a hindrance for the electron transfer to the oxidant. Simple estimates show that the redox potential of the $RH \rightleftharpoons RH^{+} + e$ process in the case of alkanes is very high and an extremely strong oxidant is required to interact with the alkanes by the mechanism of one-electron transfer to form a cation-radical. Judging by this, alkanes must be much more inert than arenes, for which the ionization potential is considerably lower.

At the same time, there is a great probability of forming neutral free radicals from alkanes in solutions of metal complexes which are strong oxidants. Thus, for example, an electron transfer with synchronous proton elimination by solvent donor molecules

$$M^{n+} + RH \xrightarrow{S} M^{(n-1)+} + R \cdot + SH^{+} \tag{1}$$

or an analogous process with the participation of ligand X^{-}

$$M^{n+}X^{-} + RH \longrightarrow M^{(n-1)+} \ldots XH + R \cdot \tag{1a}$$

are usually more favourable thermodynamically than simple electron transfer (because of the very high energy of proton solvation). According to experimental results, reactions (1) and (1a) are, in effect, a general case of alkane interactions in polar solution with strong oxidants, including metal complexes in a high oxidation state.

---

* We can conventionally define a metal complex in a medium oxidation state as one which, reacting with alkanes, behaves neither as a strong acceptor (high valency complex) nor as a strong donor (complex of low-valent metals).

Free radicals can also be formed initially by homolytic decomposition of metal complexes, in reactions of metal compounds with some intermediate products of the reactions, e.g. peroxides. The radicals formed initially will attack the alkane molecule, in the general case, to eliminate an H atom and form a hydrocarbon radical. In the presence of dioxygen the formation of radicals may result in the development of the chain reaction of alkane oxidation with metal compounds participating in the initiation and branching of the chain, propagation proceeding by means of reactions of radicals with molecules (such as $RO_2^. + RH$). Such reactions are usually of low selectivity and lead to a variety of products. However, with high concentrations of metal compounds (such as Co(III) and Mn(III)) they can take part in chain propagation, forming radicals, for example, in reaction (1) which, in some cases, results in highly selective oxidation processes which are employed industrially.

The reactions involving a radical cleavage of the C—H bond, among them reactions of type (1), are usually characterized by 'normal' selectivity towards the branched alkanes involving tertiary, secondary and primary C—H bonds: $3° > 2° > 1°$.

If we want to achieve a different selectivity and use all the other advantages of metal coordination catalysis, we should introduce the alkanes into the co-ordination sphere of a metal complex in order to turn the hydrocarbon molecule or a fragment of it into one of the ligands of the metal complex.

The interaction of alkanes with other molecules may be considered using a perturbation theory.

The formula

$$\Delta E = -\frac{q_s q_t}{R_{st}\epsilon} + \frac{2(c_s^m c_t^n \Delta\beta_{st})^2}{E_m^* - E_n^*} \tag{2}$$

expresses the energy developed on the approach of two species to each other [2]. The first term represents the electrostatic interaction between donor s and acceptor t atoms with charges q_s and q_t at distance R_{st} in the solvent with dielectric constant ϵ. The second term represents the 'orbital' interactions by the formation of covalent bond where c_s^m and c_t^n are coefficients of respective orbitals of atoms, s and t; $\Delta\beta_{st}$ is the change in the resonance integral between the interacting orbitals of atoms, s and t, at distance R_{st}; E^* are values which characterize the energy of different molecular orbitals, m and n, in isolated molecules.

Since the charges on C and H atoms in hydrocarbons are very small, the considerable energy due to Coulomb interaction may be achieved only through a large charge concentration on a species reacting with the alkane. That means it must be either a very strong hard acid or a very strong hard base. In other words, the extremely weak basic properties of alkanes require very strong acids to interact with them. In effect, alkane protonation is achieved only with the help of so-called superacids (for example, the $HF-SbF_5$ system). Olefin and arene hydrocarbons, being also rather weak bases, nevertheless react with ordinary acids, such as sulfuric acid or even aqueous solutions of acids, i.e. with H_3O^+ ions, due to the greater polarizability of unsaturated hydrocarbon molecules.

Electrophilic metal complexes in a high oxidation state can also interact with arenes and olefins by the mechanism of acidic interaction

which may result in a reaction of electrophilic proton substitution

These are the reactions of mercuration (using Hg(II)) or plumbation (using Pb(IV)) where the metal—organic compounds produced can be isolated, as well as the reactions with Pd(II) complexes, where such compounds are formed as intermediates, the final products being, for example, those of oxidative arene coupling.

The formation of positively charged intermediate species of electrophilic addition to the molecule must, naturally, facilitate the subsequent proton elimination and the formation of the substitution product. In the case of alkanes, the mechanism involving preliminary electrophile addition should be far less favourable than for olefins and arenes, since alkanes are much worse donors (the proton affinity of methane is 2 eV lower than for benzene). Hence, the reaction of electrophilic substitution for aromatic hydrocarbons cannot be generally extended to alkanes, which are expected to react only with very strong acids, including complexes of metals in high oxidation states. Although the data on high-valency metal complexes forming addition products with the alkanes, which play the role of donors, are so far virtually absent, such examples are expected to be found, taking into consideration the data on superacids.

The mechanism of acid—base interaction with alkanes becomes quite improbable for metal complexes in medium and low oxidation states. If we want them to react with alkanes, they must possess the capability for strong orbital interaction, which is expressed by the second term of equation (2). Thus the interaction of these complexes with alkanes should proceed according to the type: soft acids – soft bases [3] forming, as a general case, a metal–carbon covalent bond.

Hence, in these cases which are apparently the most interesting from the point of view of coordination catalysis, *the cleavage of* C–H (*or* C–C) *bond must proceed simultaneously with the formation of new bonds, including metal–carbon ones.*

Three mechanisms of C–H bond activation in the reactions of alkanes with transition metal complexes leading to the formation of metal–carbon bonds can be visualized as follows:

1. Formation of a metal-alkyl derivative by way of substitution of the ligand (X) without change in the metal oxidation state

$$L_nMX + RH \longrightarrow L_nM\underset{X}{\overset{R}{\diamond}}H \longrightarrow L_nM{-}R + HX \tag{3}$$

In particular cases, e.g., when X is an oxygen atom having a double bond to the metal (or a double negative charge), the HX molecule formed may remain in the coordination sphere:

$$M{=}O + RH \longrightarrow M\underset{OH}{\overset{R}{<}} \tag{3a}$$

2. Oxidative addition with formation of alkyl hydride

$$L_nM + RH \rightleftharpoons L_nM\underset{H}{\overset{R}{<}} \tag{4}$$

3. Oxidative addition with co-action of the ligand

$$L_nM \cdots L' + RH \rightleftharpoons L_nM\underset{L'H}{\overset{R}{<}} \tag{5}$$

e.g., when L' is the olefin

$$\underset{C{=}C}{\overset{L_nM}{>}}{<} + RH \rightleftharpoons L_nM\underset{C{-}C}{\overset{R}{<}}H \tag{5a}$$

Reaction (3) must be expected for complexes in a high or medium oxidation state, whereas reactions (4) and (5) must be typical for complexes of medium and, particularly, low oxidation state. Apparently reactions (3–5) require a low activation energy, provided they are thermodynamically allowed. In effect, the reverse reactions to (3–5) are known to be fast processes in the chemistry of organometallic compounds. The reaction which is a reversed reaction (3) is the well known hydrolytic cleavage of the metal—carbon bond

$$M-R + HX \longrightarrow M^+X^- + RH$$

The reaction which is to be considered as a reverse process to (4) is a reductive elimination of alkane from alkyl hydride, which is a generally accepted step in the catalytic hydrogenation of olefins

$$L_nM \xrightarrow{H_2,} L_nM \begin{array}{c} H \\ C-C \end{array} H \longrightarrow L_nM + \begin{array}{c} H \\ C - C \end{array} H$$

Finally, the formation of alkane and olefin via intramolecular disproportionation of two alkyl groups in the metal coordination sphere (a reverse process to (5a)) is a known reaction for dialkyl complexes, for example

$$L_nM \begin{array}{c} C_2H_5 \\ C_2H_5 \end{array} \longrightarrow L_nM + C_2H_6 + C_2H_4$$

According to the principle of microscopic reversibility, reactions (3–5) should proceed readily, provided the energies of the newly formed bonds compensate the energies of the broken ones. Thus, in the reactions of alkanes, the strength of this bond is a very important factor as a driving force for the reaction, since the cleavage of the C—H bond should proceed synchronously with the M—C bond formation.

At present a great number of compounds with transition metal-alkyl σ-bonds are known. According to present-day knowledge [4], early assumptions of the weakness of such bonds proved to be invalid. The reactivity of alkyl derivatives of transition metals is usually associated with their kinetic (but not thermodynamic) instability, for example, reactions of hydrogen β-elimination are common here. Table II gives energies of M—C σ-bonds (cited in [5]) recorded calorimetrically. For a number of compounds the bond energies are rather high, in the case of similar ligands the bond energy increasing when going to lower-lying periods of the Periodic System.

TABLE II

Dissociation energy of metal—hydrogen σ-bonds recorded calorimetrically

Bond	Compound	Bond energy kcal mole^{-1}
Ti—C	Ti$[CH_2C(CH_3)_3]_3$	41
Ti—C	Ti$[CH_2Si(CH_3)_3]_3$	60
Ti—C	Ti$(CH_2C_6H_5)_4$	57
Ti—C	$(\eta^5\text{-}C_5H_5)_2Ti(CH_3)_2$	60
Ti—C	$(\eta^5\text{-}C_5H_5)_2Ti(C_6H_5)_2$	84
Zr—C	Zr$[CH_2C(CH_3)_3]_4$	53
Zr—C	Zr$[CH_2Si(CH_3)_3]_4$	78
Zr—C	Zr$(CH_2C_6H_5)_4$	91
Mn—C	Mn$(CH_3)(CO)_5$	30
Re—C	Re$(CH_3)(CO)_5$	53
Pt—C	Pt$(C_6H_5)_2[P(C_2H_5)_3]$	60
Pt—C	$(\eta^5\text{-}C_5H_5)Pt(CH_3)_3$	38

Substitution mechanisms (3) and (3a) for high valency metal complexes are more thermodynamically favourable than the mechanisms of electron transfer with simultaneous radical formation (reactions (1) and (1a)) by virtue of the M—R bond energy. However, reactions (3) and (3a) must be more sterically hindered and this could be the reason why radical formation is the general mechanism of the interaction of alkanes with metal complexes in high oxidation states, and the 'normal' order of reactivity, $3° > 2° > 1°$, is generally observed.

The reversed order of reactivity, $1° > 2° > 3°$, could be expected for mechanisms (3) and (3a), since the primary C—H bond is the least sterically hindered and, although it is also the strongest one, the rise of the bond energy from the tertiary to the primary C—H bond will be more or less compensated by the respective rise of the M—R bond energies.

We shall see below that the important role of steric hindrance is, in effect, observed in some cases of biological oxidation with the participation of high valency iron complexes involved in the enzyme, and also for some nonenzymatic oxidation reactions with high valency metal complexes, particularly Co(III) ones. While mechanisms (3—3a) could be tentatively suggested for these cases, it has not yet been established unequivocally for any of the oxidation reactions.

The oxidation addition of the H_2 molecule, in the case of low valency metal complexes, is a well-established fact, for example [6],

$$Ir^I(PPh_3)_2(CO)X + H_2 \longrightarrow H_2Ir^{III}(Ph_3P)_2(CO)X$$

Comparing reaction (4) of oxidative addition for methane and dihydrogen, it may be concluded that the reaction enthalpy in the case of methane must be lower than for dihydrogen due to the difference in the M—H and M—CH$_3$ bond energies. This difference can be approximately estimated using the Pauling formula, if we neglect the differences in electronegativities of H atom and alkyl radical, $(\chi_R - \chi_H)$. In which case,

$$\Delta = D(M-CH_3) - \tfrac{1}{2}D(M-M) - \tfrac{1}{2}D(CH_3-CH_3)$$
$$\Delta = D(M-H) - \tfrac{1}{2}D(M-M) - \tfrac{1}{2}D(H-H)$$

Hence:

$$D(M-CH_3) \simeq D(M-H) - \tfrac{1}{2}[D(H-H) - D(CH_3-CH_3)]$$
$$= D(M-H) - 10 \text{ kcal mole}^{-1}.$$

Therefore, the energy of the covalent M—CH$_3$ bond should be about 10 kcal mole^{-1} less than the energy of the M—H bond.

The C—H bond energy in methane being almost equal to that of the H—H bond, it may be inferred that the oxidative addition for methane should be approximately 10 kcal less exothermic (or more endothermic) than for dihydrogen. Thus both the steric and thermodynamic factors for methane oxidative addition are less favourable than for dihydrogen. However, the difference is not so large as to believe this reaction for CH$_4$ would be improbable, at least, as an equilibrium process.

For alkanes other than methane the decrease of the C—H bond would be obviously compensated with the respective decrease of the M—C bond energy.

Therefore, for the mechanism of oxidative addition for metal complexes of medium oxidation state, various C—H bonds should be approximately equivalent and there should not be too marked a difference between the C—H bonds in aliphatic and aromatic hydrocarbons. At the same time the steric factors must here play a significant role. These very peculiarities have been observed in the recently discovered reactions of alkanes with Pt(II) complexes, in which the formation of Pt—R bonds is unequivocally established and there is no doubt as to the synchronous bond cleavage and formation. However, no stable alkylhydride complex is apparently produced, but rather the reaction, having started as the oxidative addition, then changes to the mechanism of substitution, in which the proton attacks one of the ligands of the platinum complex, i.e. the reaction proceeds as alkylation with the ligand co-action (mechanism (3)).

In the case of low-valent metal complexes the mechanism of oxidative addition to form alkyl hydride (reaction (4)) or with the ligand co-action (reaction (5)) becomes the most probable one. The hydrocarbon molecule in this case acts as an electron acceptor, which means that electronegative substituents can enhance the reaction. Taking into consideration the greater electronegativity of the sp^2-hybridized carbon atom in comparison with the sp^3-hybridized one, we may again expect the aromatic hydrocarbons to be more reactive than the alkanes.

At present a number of examples of oxidative addition with the cleavage of the C—H bond are known for such compounds, where it is activated by electronegative substituents, or the C—H bond in aromatic compounds with the participation of low-valent metal complexes. Although so far the alkane molecules are known to take part in such reactions in only a few cases, their number is gradually increasing and the field of activation of alkanes by metal complexes is constantly expanding.

The comparison of the reactions of alkanes with the reactions of other compounds with C—H bonds being very important (in order to be able to make some definite conclusions on the reaction mechanism and the possibility of alkane activation), the first chapters of this monograph give an account of reactions in which chemically activated C—H bonds take part. Then the reactions of alkanes with various compounds (except metal complexes) are reviewed, including the reactions of alkanes on the surfaces of metals and metal oxides which, as we now know, do not differ in principle from the reactions with metal complexes in homogeneous solutions. The last chapters deal with the reactions of alkanes with metal complexes which are of course of the main interest for this book.

References

1. O. A. Reutov, I. P. Beletskaja, and K. P. Butin: *C—H Acids*, Nauka, Moscow (1980).
2. G. Klopman: *Chemical Reactivity and Reaction Paths* (Ed. G. Klopman), p. 55, Wiley, New York (1974).
3. R. G. Pearson: *Science*, 151, 172 (1966).
4. G. Wilkinson: *Pure and Applied Chem.*, 30, 627 (1972).
5. G. Henrici-Olivé and S. Olivé: *Coordination and Catalysis*, Verlag Chemie, Weinheim—New York (1977).
6. L. Vaska and J. W. Diuzio: *J. Amer. Chem. Soc.*, 84, 679 (1962).

CHAPTER I

REACTIONS OF METAL COMPLEXES WITH COMPOUNDS
CONTAINING 'ACTIVATED' C—H BONDS

In this chapter we consider the reactions of C—H bond-containing compounds other than alkanes. As compared with alkanes, the C—H bonds in these compounds may be regarded as 'activated' by the presence of an aromatic ring (for carbon atoms which form part of the ring or are in the α-position to the ring), by a polar substituent, or by the proximity of the C—H bond to a metal atom of the complex when C—H is present in the ligand of the coordination compound.

Some examples are also given here for C—H bond cleavage by transition metal complexes in the neighbourhood of olefinic double bonds, which shows much similarity to the C—H bond cleavage in arenes.

Reactions of acetylenes are not considered, since their C—H bond with its acidic proton differs too much from that in alkanes.

I.1. Aromatic Hydrocarbons

The considerably higher reactivity of arenes as compared with alkanes is a well-known fact in chemistry. Aromatic molecules are far better donors and acceptors of electrons and they also possess much greater polarizability. These properties, in the case of aromatic molecules, ensure the formation of rather stable intermediates in numerous substitution reactions, for example,

Reactions of arenes in solutions containing metal complexes have also been known for a long time.

I.1.1. REACTIONS OF ARENES WITH ELECTROPHILIC OXIDANTS

Metal compounds and complexes which are strong oxidants (Co(III), Pb(IV), Mn(VII)) react with benzene and other aromatic molecules to form phenols or their derivatives as intermediates. Acidic media and ligands with strong acceptor properties favour the reaction; for example, cobalt(III) tristrifluoroacetate oxidizes benzene rapidly (30 min) in trifluoroacetic acid with 90% yield at $25°C$ (this reaction proceeds at a much higher rate than with cyclohexane) [1, 2].

It is possible that oxidants which are also strong electrophiles react with arenes initially via the mechanism of electrophilic substitution. In certain cases this mechanism has been confirmed by the observance of aryl metal derivatives.

I.1.2. MERCURATION AND SIMILAR REACTIONS

The mercuration reaction [3, 4] is the most intensively studied reaction among those in which the C—H bond is replaced by the C—M one. The reaction follows the scheme:

$$ArH + HgX_2 \rightleftharpoons ArHgX + HX; \qquad X = NO_3, ClO_4, OAc.$$

This reaction is employed to synthesize various mercury-organic compounds.

Mercuration is a reversible process, and the equilibrium can be shifted to the left by adding acids. The reaction is usually carried out by mercuric acetate or perchlorate in solutions of acetic or aqueous perchloric acid. The reaction rate is enhanced with the increase of perchloric acid concentration. The effect of acid might possibly be caused by formation of more electrophilic species, for example, $AcOHg^+$ or $AcOHgO^+(H)Ac$. A considerable primary isotope effect is observed in mercuration [5, 6]. Thus in the reaction of benzene with mercuric acetate in acetic acid $k_H/k_D = 6.0 \pm 0.1$ [6], which demonstrates a strong participation of the C—H bond cleavage in the rate-determining step.

Mercuration of aromatic molecules reveals many features of a typically electrophilic substitution: electron-donating substituents accelerate the reaction while electron-withdrawing ones slow it down. The aromatic amines are particularly readily mercurated (in aqueous solutions at room temperature); phenols also possess high reactivity; while nitrobenzene and pyridine are mercurated with more difficulty than is benzene.

The entry of mercury into aromatic compounds obeys the rules of electrophilic substitution: o-, p-orientation for electron-donating substituents and m-orientation for electron-withdrawing ones (Table I.1) [7]. For substituted benzenes a very good correlation is found between the relative rate data with the substituent constant σ^+. Thus the modified Hammett equation, $\log k_s/k_0 = \rho\sigma^+$, is applicable with reaction constant ρ calculated to be -4.00.

Some steric hindrance for mercuration in the o-position takes place (being particularly strong for such substituents as t-butyl and phenyl). However, they are almost insignificant in the case of other, less bulky, substituents.

TABLE I.1

Partial rate factors in noncatalytic mercuration of monosubstituted benzenes, C_6H_5X, using Hg $(OAc)_2$ at 25°C in glacial acetic acid

X	Relative rate	Isomer distribution (%)			Partial rate factors		
		ortho	meta	para	ortho	meta	para
OMe	448	14.0	0	86.0	188		2310
AcNH	46.1			(100)[a]			277
OPh	64.8			(100)[a]			194
Me	6.54	29.4	11.5	59.1	5.77	2.26	23.2
t-Bu	4.00	0.0	28.4	71.6		3.41	17.2
Ph	2.71	2.0	19	79	0.081	0.77	6.42
H[b]	1.00				1	1	1
F	0.0702	28.7	1.8	69.5	0.60	0.038	2.92
Cl	0.100	25.1	25.8	46.0	0.075	0.054	0.34
Br	0.090	26.6	22.4	51.0	0.072	0.060	0.275

[a] 100% substitution is accepted in the p-position;
[b] rate constant for benzene at 25°: 0.0413×10^{-6} $M^{-1}s^{-1}$;
    $\Delta H^{\neq} = 21.6$ kcal mole^{-1}; $\Delta S^{\neq} = 19.6$ e.u.

A somewhat preferential formation of o-isomers is observed in some cases of mercuration by mercuric acetate in nonpolar media (for nitrobenzene, benzoic acid, benzophenone). This is evidently explained by formation of mercuric complexes with oxygen atoms in substituents, which locate mercuric ions close to the o-position of the benzene ring.

It is essential that the isomer distribution changes during the course of reaction; for example, the amount of m-isomer increases with time in the case of an o-p-orienting substituent. This might be explained by the reversibility of mercuration reactions, which makes it possible to isomerize the substituted products obtained, followed by a build-up of the thermodynamically most stable product during the course of the reaction. The reaction in these cases proceeds at rather high temperatures (up to 150°C).

If we take a more ionized mercuric salt, for example mercuric perchlorate in aqueous perchloric acid, the isomer distribution in mercuration obeys the rules of normal electrophilic substitution.

Mercuration of aromatic compounds is an example of an interaction in which the electrophile is a rather strong soft acid. The stability of the Hg—C bond combined with the electrophilic properties of mercuric salts interacting with such comparatively strong nucleophiles as arenes is the driving force of the process. Obviously the alkane reactions do not occur under similar conditions.

It should be mentioned that reactions of thallation, plumbation and auration are likely to proceed in a similar manner to mercuation [8–10]; compounds containing such fragments as Pb—Ar, Tl—Ar, and Au—Ar have been isolated in reactions of benzene derivatives with electron-donating substituents, e.g.,

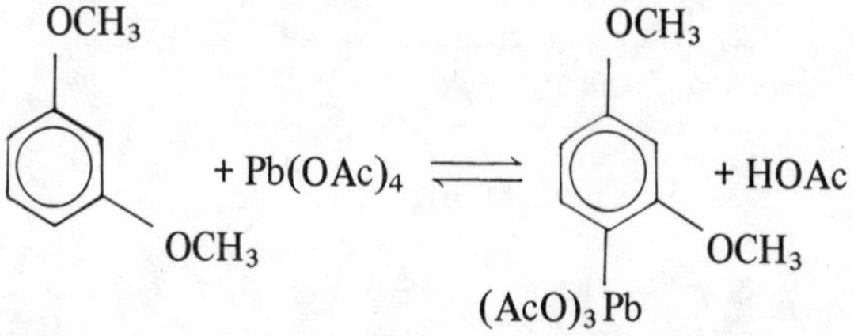

I.1.3. OXIDATION OF AROMATIC MOLECULES BY BIVALENT PALLADIUM

Palladium(II) salts give rise to several oxidation reactions of aromatic hydrocarbons with the cleavage of the C—H bond of the ring. They are:

1. Oxidative coupling of aromatic molecules:

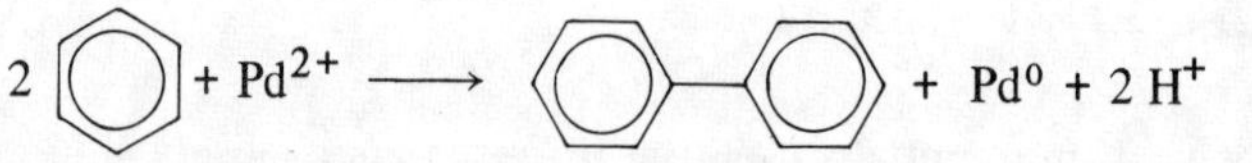

$$2 \bigcirc + Pd^{2+} \longrightarrow \bigcirc-\bigcirc + Pd^0 + 2 H^+ \qquad I.1$$

2. Oxidative coupling of arenes and olefins:

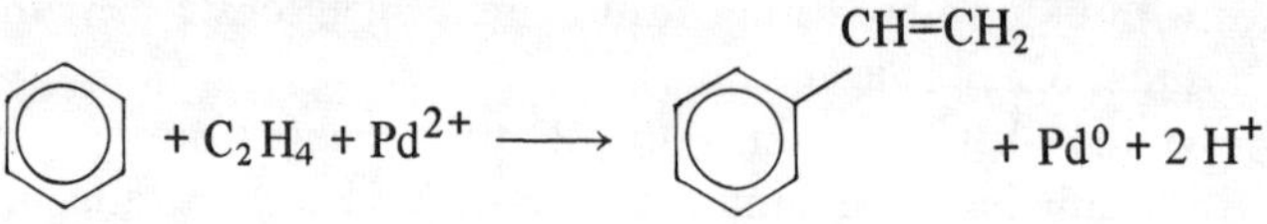

3. Acetoxylation:

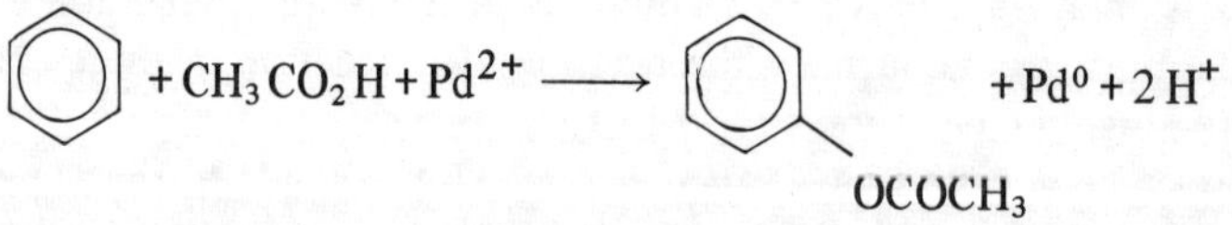

These reactions have attracted much attention and have been reviewed in a number of articles [11]. The reaction of oxidative coupling was first described in 1965 by Van Helden and Verberg [12] who showed that it occurs in acetic acid. Later on it was shown that the yield and composition of products depend on the reaction conditions. The yield of diaryls changes in the range of 25–81% depending on the substrate in reactions of arenes with $PdCl_2$ + NaOAc in acetic acid. Without sodium acetate the reaction does not occur. In the cases of $PdBr_2$ and PdI_2 the reaction does not occur even in the presence of sodium acetate.

Alkylbenzenes yield diarylmethanes in the oxidative coupling reaction, besides diaryls; moreover, only diarylmethanes are produced from such compounds as mesitylene and durene, the mechanism of these reactions being evidently different from that of diaryl formation.

High yields of diaryl are observed when using palladium complexes with olefins in the oxidative coupling of arenes. Thus diphenyl is produced with a quantitative yield (1 : 1) per molecule of palladium compound taken in the case of the $C_2H_4 \cdot PdCl_2$ -$AgNO_3$ system. Silver nitrate itself is inactive in the oxidative coupling reaction.

The oxidative coupling of arenes reveals the properties of electrophilic substitution in the aromatic ring: electropositive substituents accelerate the reaction, and electronegative ones retard it. Monosubstituted arenes give all the six possible isomers in oxidative coupling, though the content of 2,2-substituted product is usually low due to steric hindrance. In the case of monoalkyl-substituted benzene the main products are 3,4- and 4,4-substituted diaryls.

Oxidative coupling does not occur for such substituted arenes as p-diisopropylbenzene, p-di(t-butyl)benzene and mesitylene, obviously due to steric hindrance (the formation of diarylmethanes in the case of mesitylene, as was mentioned above, proceeds by a different mechanism).

The kinetics of benzene oxidative addition in the $PdCl_2$ -NaOAc system in acetic acid follows the equation

$$v = k\,[Pd^{II}]\,[C_6H_6]$$

and although the presence of sodium acetate is necessary, its concentration is not present in the kinetic equation.

A large isotope effect (k_H/k_D = 5.0) is observed in the oxidative coupling, which is close to that in arene mercuration, thus supporting the mechanism of electrophilic substitution as the rate-determining step. Another significant support of σ-aryl palladium derivatives formation in the first stage of the reaction is the observation of diaryls forming in the reactions of σ-aryl metal derivatives (Hg(II), Tl(III), B(III), etc.) with palladium(II) complexes.

In this case the first stage obviously produces an arylpalladium derivative

$$ArHgX + PdX_2 \longrightarrow ArPdX + HgX_2$$

which gives an oxidative coupling product

$$2\ ArPdX \longrightarrow Ar—Ar + PdX_2 + Pd$$

For diarylmercury compounds (e.g., (p-tolyl)$_2$ Hg) the oxidative coupling occurs instantaneously.

Therefore, the first stage, determining the rate of arene oxidative coupling under the action of palladium complexes, is presumably the electrophilic substitution of the proton

The role of acetate in the coordination sphere of palladium consists of facilitating the proton elimination.

The substitution reaction is obviously irreversible, since no H–D exchange occurs with the medium.

The diaryl formation in Reaction (I.1) is evidently preceded by disproportionation of alkylpalladium derivatives

$$2 \, ArPdX \rightleftharpoons Ar_2Pd + PdX_2$$

and the coupling of two aryl groups proceeds in the coordination sphere of the palladium complex. It should be noted that this mechanism of diaryl formation is apparently a common one in solutions of aryl derivatives of transition metals.

The reaction of olefin arylation by bivalent palladium complexes, as well as arene acetoxylation, might also involve electrophilic proton substitution in the aromatic ring under the action of Pd(II). The further interaction, in the case of olefin, proceeds by means of an olefin insertion in the Pd—Ar bond.

$$
\begin{array}{c}
CH_2 \\
\| \longrightarrow Pd{-}Ar \longrightarrow Pd{-}CHR{-}CH_2Ar \longrightarrow ArCH{=}CHR \\
CH \\
| \\
R
\end{array}
$$

An alternative mechanism involving the σ-vinyl Pd complex seems less probable. Thus the isotope effect, when H is replaced by D, in benzene $k_H/k_D = 5.0$, whereas for deuterated olefin $k_{C_2H_4}/k_{C_2D_4}$ reaches only 1.2 ± 0.2 [13].

It is worth noting that the oxidative reactions of aromatic compounds in the presence of palladium complexes are, for the most part, analogous to the well-known catalytic olefin oxidation, for example,

$$C_2H_4 + \tfrac{1}{2}O_2 \xrightarrow{\ PdCl_2\ } CH_3CHO$$

$$C_2H_4 + HOAc + \tfrac{1}{2}O_2 \xrightarrow{\ PdL_2\ } CH_2{=}CHOAc$$

(see, e.g., [14]).

Oxidative dimerization of styrene or 1,1-diphenylethylene in aqueous acetic acid is similar to the arene reactions with participation of Pd^{2+}

$$2 Ph_2C{=}CH_2 + PdX_2 \longrightarrow Ph_2C{=}CH{-}CH{=}CPh_2 + 2 Pd^0 + 2 HX$$

The reaction evidently involves the electrophilic proton replacement by Pd in the CH_2 group of the olefin

$$Ph_2C{=}CH_2 + PdX_2 \longrightarrow Ph_2C{=}CHPdX + H^+ + X^-$$

The subsequent process with the olefin proceeds analogously to the oxidative coupling of olefins with arenes [15, 16].

I.1.4. REACTIONS OF ARENES IN PLATINUM SALT SOLUTIONS

In 1967 Garnett and Hodges reported on the catalysis by Pt(II) complexes of the homogeneous H—D exchange between benzene and a mixture of D_2O + CH_3COOD as solvent [17]. Besides the H atoms of the aromatic ring, those in the alkyl groups also take part in the alkylbenzene exchange, among them H atoms at the remote side-chain carbon atoms.

Oxidation of aromatic compounds by the Pt(IV) complex, H_2PtCl_6, to produce chlorosubstituted arenes and small amounts of biaryls is also catalyzed by Pt(II) salts.

It was further shown elsewhere that the reaction involves intermediate formation of Pt(II) and Pt(IV) aryl derivatives. Unexpectedly, the Pt(IV) aryl derivatives turned out to be very stable at comparatively high temperatures and in such agressive media as mixtures of water and trifluoroacetic acid. This makes it possible not only to observe them by means of NMR techniques but also to isolate them from solution in crystalline form. The Pt(II) complex-catalyzed hydrocarbon reactions appeared to be common to both the aromatic and aliphatic hydrocarbons, the platinum complex-catalyzed reactions of alkanes presenting the first examples of activation of alkanes by metal complexes in solution.

Reactions of arenes in Pt(II) complex-containing solutions, along with the experimental results on the reactions of alkanes, will be more extensively described in Chapter V.

I.1.5. REACTIONS OF ARENES WITH LOW OXIDATION STATE METAL COMPLEXES

Arenes often react with electron-rich metal complexes forming arylmetal hydrides, thus oxidative addition with the increase of metal oxidation state is observed (see the review [18]).

Chatt and Davidson were the first to demonstrate such a possibility for the case of the ruthenium complex, $RuCl_2(dmpe)_2$ [19]. During the reaction of this with metallic sodium, a highly reactive Ru^0 species is evidently formed which further reacts with aromatic molecules. Thus β-naphthylruthenium hydride is formed with naphthalene.

$$RuCl_2(dmpe)_2 \xrightarrow{Na} Ru^0(dmpe)_2 \xrightarrow{C_{10}H_8}$$

where dmpe $= (CH_3)_2 PCH_2 CH_2 P(CH_3)_2$. An analogous osmium complex reacts in the same way.

In the case of the similar iron complex $(FeCl_2(dmpe)_2)$ its formation and further reaction with naphthalene leads again to the formation of a β-naphthyl derivative, which may be used to study the reactions with other arenes, the naphthalene reaction being reversible [20].

$$\underset{}{} Fe(dmpe)_2 \rightleftharpoons Fe^0(dmpe)_2 \xrightarrow{C_6H_6}$$

Meta- and *p*-tolyl derivatives are obtained in the reaction with toluene.

In the absence of arenes the ruthenium complex interacts with C—H bonds of methyl groups of dmpe (see below).

In the cases considered, the formation of the reaction products in the reaction with alkanes is unexpected, though the complex reaction with alkanes might proceed with a measurable rate, comparable to that of the reaction with phosphine methyl groups. The alkylhydride formation reaction having an equilibrium character, the reaction with an alkane is presumably shifted to the left, even in the absence of ligand interaction with C—H bonds. The competition with a parallel reaction involving a methyl group from phosphine with a more favourable entropy factor almost completely excludes the formation of even small amounts of the alkylhydride complex from the alkane.

The oxidative addition of aromatic molecules is observed in the reaction of dicyclopentadienyltungsten $(Cp_2 W)$ formed as an intermediate species, e.g. in the photolysis of $Cp_2 WH_2$ [21a] or in its thermal reaction with dienes [21b]

$$Cp_2 WH_2 \xrightarrow{h\nu} Cp_2 W \xrightarrow{C_6H_6} Cp_2 W \overset{H}{\underset{Ph}{<}}$$

where Cp $= \pi\text{-}C_5 H_5$.

Such a reaction does not occur with tungstenocene analogs, Cp_2Cr and Cp_2Mo. The absence of interaction products in the case of alkanes in the presence of Cp_2W does not really mean that the reaction rate is negligible: the possibility of the competing reaction with the C—H bond of the cyclopentadienyl groups of tungstenocene derivatives should be also taken into consideration.

Cleavage of the C—H bond in olefins, and formation of metal-carbon σ-bonds, have been reported for platinum, rhodium and iridium complexes [22a, b]. Oxidative addition involving vinylic C—H bond cleavage has also been reported for the reaction of alkyl methacrylate with dihydridotetrakis(triphenylphosphine)ruthenium(II) in which the structure of the resulting complex having a ruthenium-carbon σ-bond has been determined by X-ray structural analysis [22c].

I.1.6. ISOTOPE EXCHANGE OF ARENES WITH DEUTERIUM

The complexes able to form tri- and polyhydrides with dihydrogen can catalyze H—D exchange of aromatic hydrocarbons with deuterium by the following mechanism (see reviews [18, 23]):

$$MH_3 \underset{-H_2}{\rightleftarrows} MH \underset{+RH}{\rightleftarrows} M{\overset{R}{\underset{H}{\diagdown}}}H \underset{-H_2}{\rightleftarrows} M-R \underset{+D_2}{\rightleftarrows} M{\overset{R}{\underset{D}{\diagdown}}}D \underset{-RD}{\rightleftarrows} MD$$

Such catalysis may occur with cyclopentadienyl derivatives of tantalum and niobium, Cp_2TaH_3 and Cp_2NbH_3. It is interesting to note that, with the tantalum complex, the reaction rate is practically insensitive to polar substituents in the benzene ring (to both donor and acceptor) and, with the niobium complex, acceptor substituents (CF_3 and F) enhance the reaction rate to some extent. Therefore, in the oxidative addition in this case the donor and acceptor properties of the tantalum complex are practically equal. Alkanes and even activated C—H bonds in the side chain of the benzene ring are inactive in these systems. This can be explained by the low stability of the Nb—C and Ta—C bonds in the case of aliphatic groups in comparison with aromatic ones. The alternative explanations might be connected with the detailed reaction mechanism. For example, the arene might at first displace H_2 from the coordination sphere forming a π-arene complex and only then interact along the C—H bond. Naturally, in this case the alkanes should be much less reactive than arenes.

The H—D exchange can be also initiated by other hydride derivatives: $TaH_5(Me_2PCH_2CH_2PMe_2)$, $ReH_5(PPh_3)_3$, $IrH_5(PMe_3)_2$.

Some complexes can exchange H atoms between arenes and olefins. The H—D exchange occurs in the $CpRh(C_2H_4)_2$ in the presence of deuterated benzene-d_6 both in the ethylene molecule and in the C_5H_5 group, evidently, via the following mechanism [24]:

$$CpRh(C_2H_4)_2 \underset{-C_2H_4}{\rightleftharpoons} CpRh(C_2H_4) \underset{+C_6D_6}{\rightleftharpoons} CpRh-C_6D_5 \rightleftharpoons$$

with structures:

$$CpRh \Big\langle \begin{matrix} C_6D_5 \\ CH_2-CH_2D \end{matrix} \rightleftharpoons CpRh \Big\langle \begin{matrix} C_6D_5 \\ H \end{matrix} \quad \text{etc.}$$

I.2. Reactions of 'Activated' Aliphatic C—H Bonds

The aromatic ring activates the C—H bond in the alkyl group in the α-position to the ring. The C—H bond dissociation energy in alkylaromatic compounds is much lower than the C—H bond energy in alkanes (e.g., in the toluene CH_3-group it is 18 kcal mole^{-1} lower than in methane). The ionization potential, characteristically for aromatic compounds, is much lower, whereas the acidity of the σ-C—H bond in the α-position in alkylaromatic compounds is higher than in alkanes, which makes alkylaromatic compounds more reactive than alkanes towards free radicals, electrophiles and nucleophiles. Reactions of alkylaromatic compounds with metal complexes involve the intermediate formation of a M—C bond which can be stabilized by interacting with the aromatic ring (interaction of the π-benzyl type).

Owing to these characteristics of alkylaromatics, the reactions of α-C—H bonds with various active particles (metal compounds and complexes among them) are often confined to these bonds exclusively and have not been observed for non-activated C—H bonds. In certain cases, however, only qualitative differences exist.

I.2.1. C—H BONDS IN THE α-POSITION OF THE SIDE CHAIN OF THE AROMATIC RING: REACTIONS WITH OXIDANTS

Electrophilic oxidants can react with alkylbenzenes to produce free radicals [2]. The formation of radicals or atoms in the presence of the oxidant can be followed by the hydrogen atom abstraction from the α-position of the benzene ring

$$R \cdot + C_6H_5CH_3 \longrightarrow RH + C_6H_5CH_2 \cdot$$

This mechanism operates in reactions which involve Mn(III) compounds (with subsequent homolysis) or Co(III) compounds in the presence of Br^-:

$$Co^{3+} + Br^- \longrightarrow Co^{2+} + Br \cdot$$

$$Br \cdot + C_6H_5CH_3 \longrightarrow HBr + C_6H_5CH_2 \cdot$$

The benzyl radicals can then be oxidized:

$$C_5H_5CH_2 \cdot + M^{n+} \longrightarrow C_6H_5CH_2^+ + M^{(n-1)+}$$

In the presence of an electron acceptor, electron transfer can take place with simultaneous deprotonation of an organic molecule:

$$M^{n+} + C_6H_5CH_3 \longrightarrow C_6H_5CH_2 \cdot + M^{(n-1)+} + H^+$$

Here, as in the above case, primary products are stable benzyl radicals which enter subsequently into oxidation reactions.

Finally, with strong electron acceptors, cation-radicals can be produced at the initial stage. This mechanism is most typical for aromatic molecules with low ionization potentials ($IP < 8$ eV)

when they react with a strong oxidant, particularly in acid media. For instance, there is strong reason to believe that in Co(III) salt solutions the primary process which involves alkylaromatic compounds is, essentially, the electron transfer. Trifluoroacetic acid enhances the oxidative properties of cobalt (III) acetate. The oxidative reaction proceeds selectively at 25°C

$$ArCH_3 + Co(OAc)_3$$

$$\xrightarrow[\text{N}_2]{\text{TFA-HOAc}} ArCH_2OAc$$

$$\xrightarrow[\text{O}_2]{} ArCO_2H$$

Addition of LiCl was found to promote the oxidation of toluene, and what is more, at high Cl^- concentrations chlorination of the side chain is accompanied by the chlorination of the benzene ring, possibly as a result of the primary formation of the cation-radical [25]. The following scheme is in qualitative agreement with the dependence of toluene yield on LiCl concentration given in Table I.2.

TABLE I.2

Relative yields of chlorotoluene and chlorobenzene upon toluene oxidation by $Co(OAc)_3$ in CF_3COOH

[LiCl] (M)	Product yield	
	$ClC_6H_4CH_3$	$C_6H_5CH_2Cl$
2.1	32	46
0.73	0.7	78
0.24	–	70

It may be noted that selective catalytic oxidation of alkylaromatic compounds has important practical uses as, for instance, in the production of terephthalic acid from p-xylene [15]. The oxidation is performed by air at $225°C$ and 15 atm, with a mixture of Co(II) acetate, Mn(II) acetate and sodium bromide as a catalyst. The high-purity (99.96%) terephthalic acid is reported to have been obtained selectively. The hydrocarbon oxidation mechanism will be dealt with in more detail in Section IV.2.

I.2.2. REACTIONS OF ALKYLAROMATIC COMPOUNDS IN THE PRESENCE OF PLATINUM(II) COMPLEXES

In the presence of $PtCl_4^{2-}$ the C—H bond in the side chain of a benzene ring interacts with the solvent (D_2O and $D_2O + CH_3COOD$) to affect H—D exchange. Also, in the presence of $PtCl_4^{2-}$, the side chain of the benzene ring gets oxidized under the action of H_2PtCl_6. These reactions are similar to C—H bond reactions in an aromatic ring and alkanes. They will be considered in more detail in Chapter V.

I.2.3. REACTIONS OF TOLUENE IN THE PRESENCE OF PALLADIUM COMPLEXES

The Pd(II) compounds present in acetic acid solutions oxidize toluene to form oxidative coupling products (with the C—H bond of an aromatic ring activated) and to affect oxidative acetoxylation into the side chain to produce benzyl acetate. The latter reaction occurs with excess sodium acetate, and the benzyl acetate yield may be as high as 92%. This reaction can be made catalytic by supporting a catalyst on a silica gel or activated charcoal and using air as an oxidizing agent. The two processes — oxidative coupling and acetoxylation — were earlier thought to have occurred under the action of Pd(II). As was shown recently, the reaction in the ring seems to proceed initially as an electrophilic substitution of H^+ due to Pd^{2+}. Such a reaction in the side chain of the benzene ring seems to be unlikely. Kinetic curves for benzene acetate accumulation show that there is an induction period [26]. The addition of palladium black promotes the reaction. Acetoxylation, studied with phenanthroline-stabilized carbonyl acetate complex of palladium(I), $CH_3COOPd(CO)_4 \cdot 2AcOH$, as starting material, has enabled the authors [26] to conclude that the active complex in acetoxylation is a cluster complex which contains several Pd(0) atoms. Intermediate formation of the Pd—H bond is supported by the observation of H—D exchange between the methylene group of $C_6D_5CD_3$ and CH_3COOH on the same catalyst. It has been proposed that hydrogen atom of the CH_3 group in the coordinated toluene [26] is transferred to one of the palladium atoms (to react with the activated O_2 molecule) while the coordinated fragment

$C_6H_5CH_2$ undergoes nucleophilic attack by acetate or acetic acid. This mechanism appears to be similar to that of the heterogeneous catalytic reaction caused by palladium black.

The C—H bond in the aromatic ring side chain can also become activated by interacting with a mononuclear complex of a metal in low oxidation state, possibly by oxidative addition. For instance, the mesitylene reaction with Cp_2W produces a complex which contains two benzyl-type groups linked to the tungsten atom

The structure of the complex has been identified by X-ray analysis [27].

I.2.4. ACTIVATION OF C—H BONDS BY POLAR SUBSTITUENTS

Electronegative substituents, when introduced into alkane molecules, may cause the decrease of the oxidation rate in reactions which occur by either electron transfer (from RH to an oxidant) or proton electrophilic substitution mechanism. For instance, permanganate, while reacting in 80% trifluoroacetic acid with the alkanes of methane series, remains totally inactive under the same conditions towards nitroethanes, propyl nitrile, and propanoic acid [28]. This effect of polar substituents potentially provides the means to stop alkane reactions with strong electrophilic oxidants at the initial oxidation stages (e.g., in alcohol formation). This might be responsible for selective oxidation in biological systems (Section IV.2.4.).

Conversely, in reactions which involve complexes with metals in low oxidation state having strong electron-donor properties, the electronegative substituents can facilitate oxidative addition via the C—H bond cleavage and stabilize the higher oxidation state being formed [18]. Thus, the Fe(0) complex formed in the decomposition of naphthyl hydride reacts with acetonitrile [20]

$$Fe^0(dmpe)_2 + CH_3CN \longrightarrow \underset{NCCH_2}{\overset{H}{\diagdown}} Fe(dm\mu e)_2$$

In addition to CH_3CN, acetone, ethyl acetate and methyl cyanoacetate are found to take part in similar reactions. The Iridium(I) complex, $Ir(dmpe)_2^+$, also causes the oxidation addition of acetonitrile. Although this reaction is reversible, the complex can be identified by NMR

$$Ir(dmpe)_2^+ + CH_3CN \rightleftharpoons \underset{CH_2CN}{\overset{H}{\underset{|}{\overset{|}{Ir^+(dmpe)_2}}}}$$

The equilibrium can be shifted to the right in the presence of CO_2 in which case the resulting product, a cyanoacetate, can be isolated

$$\underset{NCCH_2}{\overset{H}{\diagdown}} Ir^+(dmpe)_2 \xrightarrow{CO_2} NCCH_2 \longrightarrow \underset{\overset{||}{O}}{C} \longrightarrow O \overset{H}{\diagdown} Ir^+(dmpe)_2$$

The reaction of acetonitrile with an iridium(0) complex produces an Ir(I) hydride and a cyanomethyl radical [29].

The oxidative addition of C–H bonds of substituted methanes activated by electronegative substituents is believed to occur as an intermediate stage in certain reactions. For instance, Cp_2WH_2, when exposed to light, reacts with methanol to form a methyl derivative, possibly by the following mechanism [30]:

$$Cp_2WH_2 \xrightarrow[-H_2]{h\nu} Cp_2W \xrightarrow{MeOH} \underset{H}{\overset{HOCH_2}{\diagdown}} WCp_2W \xrightarrow{-OH^-}$$

$$\left[\underset{H}{\overset{CH_2}{\diagdown}} WCp_2 \right]^+ \rightleftharpoons [CH_3WCp_2]^+ \xrightarrow{CH_3O^-} \underset{CH_3O}{\overset{CH_3}{\diagdown}} WCp_2$$

Irradiation of solutions of Cp_2WH_2 in tetramethylsilane gives the *cis*- and *trans*-isomers of the reaction product [31]:

and

The authors of [31] propose that the reaction proceeds by initial insertion into the C—H bond of tetramethylsilane giving $Cp_2WH(CH_2SiMe_3)$ which then produces final products by several steps including a second photoinduced reaction.

Thus tungstenocene proves to be capable of insertion into apparently almost 'non-activated' sp^3 C—H bond of tetramethylsilane.

A similar approach in the case of dihydridoiridium complexes leads to the observation of insertion of coordinatively unsaturated iridium species into the C—H bond of alkanes (see Section V.3.3).

The oxidative addition of nitromethane as a result of interaction with a Pt(0) complex has been proposed in [32].

I.2.5. ACTIVATION OF C—H BONDS IN LIGANDS OF METAL COMPLEXES

If a ligand in a metal complex contains a C—H bond which is closely adjacent to the metal atom in some of the possible conformations, the conditions become most favourable for the interaction (e.g., oxidative addition) to take place. The spatial proximity ensures both enhanced reaction rates and increased thermodynamic stability of the metallacycle produced. It can be inferred, from a reaction known as cyclometallation, e.g., that the metal in a given oxidation state and surrounded by the appropriate ligands is, in principle, capable of reacting with alkanes under more drastic conditions, though the complex which undergoes cyclometallation does not seem to be particularly suitable for the alkane-involving reaction itself, since the intramolecular cyclometallation — even in the case of a 'non-activated' C—H bond similar to that in alkanes — will proceed much faster than the reaction with the alkane. Cyclometallation is a specific case of a variety of cyclization reactions which can proceed several orders of magnitude faster than similar intermolecular reactions and, notably, the resulting cycle can remain reasonably stable even where the equilibrium in the respective intermolecular reaction is shifted towards the decomposition products.

Naturally, when the C—H bond in a ligand is additionally activated by some electronic factors (aromatic C—H bond, C—H bond in the α-position to the benzene ring, polar group-activated C—H bond), formation of metallacycles proceeds particularly readily.

ortho-Metallation is a good example. Here, the reaction involves the C—H bond of benzene ring in the *ortho*-position to the substituent which affects binding with the metal through the substituent's N, P, O, or S atom. As the *ortho*-metallation mechanism has been treated in several works [18, 33—35], only few examples from Parshall's excellent review [18] will be given here, starting with palladation of azobenzene

The reaction mechanism, which involves an electrophilic substitution of H^+ under the action of Pd(II), is facilitated by precoordination of the azobenzene N-atom to the palladium atom. The substitution takes place at the benzene ring more distant from the Pd atom, showing that a five-membered ring is more stable than a four-membered one. The catalysts in the reaction are bases, such as acetate, hydroxyl ions and triethylamine, which facilitate the proton withdrawal [36]. Precomplexing with the azobenzene was found to increase the rate of the reaction between Pd and aromatic ring (as compared with the reaction rates of $PdCl_2$ with benzene, under the same conditions) by 4×10^4 at 1 M concentration

[36], which is typical of the chelate effect and is to be expected. This is only one of a number of similar reactions which almost invariably give rise to a five-membered ring, provided that the nitrogen atom is a complexing agent. With low-valent metals in the complex, the reaction may have some of the characteristics of a nucleophilic reaction. Thus, in the reaction of *m*-fluoroazobenzene with $CH_3Mn(CO)_5$ which proceeds according to the scheme

it is the ring with the F atom that is being attacked; while, for electrophilic palladium cyclometallation, the proton substitution proceeds in a non-substituted benzene ring.

ortho-Metallation of aryl group-bearing phosphines commonly yields a four-membered ring, as for example, does the reaction with the Ir(I) complex:

which is a typical example of oxidative addition involving a change from d^8 to d^6 electron configuration.

Reactions of this type are often reversible

and can be detected by the H–D isotope exchange of ligand C—H bonds with molecular deuterium. *ortho*-Metallation in arylphosphite complexes results in a five-membered ring, more stable than the four-membered ring produced in the case of phosphines.

$$\text{CoH[P(OPh)}_3]_4 \rightleftharpoons \text{\large{\bigcirc}} \text{---} \underset{\text{O---P(OPh)}_2}{\overset{|}{\text{Co}}} \text{[P(OPh)}_3]_3 + \text{H}_2$$

The aliphatic C—H bond cleavage has long been known for β-elimination of hydrogen atom in metal-alkyls

$$\text{M---CH}_2\text{---CH}_3 \rightleftharpoons \underset{\text{H}}{\overset{|}{\text{M}}} \leftarrow \underset{\text{CH}_2}{\overset{\text{CH}_2}{\|}}$$

This reaction, like the reverse process of the olefin insertion into the M—H bond, participates in many important catalytic processes.

Less common is the α-elimination reaction which produces a carbene complex:

$$\text{M---CH}_3 \rightleftharpoons \underset{\text{H}}{\overset{|}{\text{M}}} \cdots \text{CH}_2$$

It has been observed for tantalum and tungsten methyl complexes.

Of greater interest, from the point of view of the analogy with alkanes, are reactions in which the reacting C—H bond is separated from the metal atom of the metal complex by several σ-bonds, and is 'activated' only by the entropy factor of proximity. Among these are cyclization reactions of platinum complexes, $\text{PtX}_2(\text{PR}_3)_2$, studied by Shaw et al. [37, 38]

The reaction does not seem to occur with small alkyl groups, e.g., triethylphosphine, while being promoted by bulky phosphines which contain, for instance, t-butyl and o-tolyl groups

$$\text{PtCl}_2(\text{PBu}^t\text{Pr}_2^n)_2 \xrightarrow{\text{LiBr}} \underset{\text{Bu}^t\text{Pr}^n\text{P}}{\overset{\text{Br}}{\diagdown}} \text{Pt} \overset{\text{PBu}^t\text{Pr}_2^n}{\diagup}$$

The bulky groups seem to prolong the life of a conformation in which the C—H bond is in direct contact with the metal atom, and also enhance the stability of the metallacycle produced.

Reversible formation of metallacycles even in a small concentration, can manifest itself in isotope H–D exchange of the ligand alkyl groups with D_2O or D_2. The findings available on several such systems indicate that five-membered cycles are again the most favourable. Thus, in the $PtCl_2$ complex with tri-*n*-propylphosphine, the H–D exchange with D_2O is confined primarily to the methyl group of propyl [39]:

whereas, in *n*-butylphosphines, the CH_3 group exchange is negligible.

In the H–D exchange of diphenyl-*n*-propylphosphine with D_2 occurring in the presence of $RuHCl(Ph_3P)_3$, the CH_3 groups of the phosphine-ligand propyls were also the first to have undergone exchange [23].

The K_2PtCl_4-catalyzed exchange of olefins, $RCMe_2CH{=}CH_2$, with D_2O and CH_3COOD proceeds selectively into the C-5 position [40]. The olefin appears to form a complex with Pt(II), and the observed selective exchange again demonstrates that five-membered cycles are normally favoured over six- and four-membered ones. The first observation of direct oxidative addition to involve C–H bond cleavage in a phosphine ligand was made by Chatt and Davidson [19] for the thermal decomposition of a ruthenium naphthylhydride complex *in vacuo*. The resulting intermediate coordinatively-unsaturated Ru(0) complex reacts with the phosphine C–H bond:

X-ray analysis identified this as a dimer [41], but a possible route could involve initial monomolecular cyclometallation in the solution to be followed by dimerization of the product. Other complexes with the d^8 electronic configuration react in a similar manner. Thus, the removal of naphthalene from the β-naphthyl hydride complex of iron yields a product similar to the one formed with a ruthenium complex [20].

The reaction of an Fe(0) complex with trimethylphosphine produces a hydride

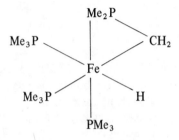

which contains a three-membered cycle [42].

A similar reaction involving the $[Ir(dmpe)_2]^+$ complex results in equilibrium and the product can be stabilized by introducing CO_2 [18]:

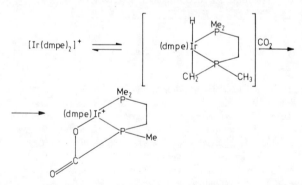

There is growing evidence that, in many cases, the C—H bond in a ligand of a metal complex approaches the metal atom to give a short M . . . H distance, which indicates positive interaction between metal and hydrogen atoms. Thus, X-ray studies show that the interaction of vinyl hydrogen with Pd in the $PdBr(PPh_3)_2(CCO_2Me)_4H$ complex leads to a Pd . . . H distance of 2.3 Å [43], whereas the sum of the Van der Waals radii is 3.1 Å. Similarly, a rather short distance (2.55 Å) between titanium and one of the methyl hydrogen atoms is reported in (2,6-dimethylphenyl)dicyclopentadienyl titanium [44].

In the alkylpyrazoleborate complexes of Ni and Mo

the methylene C—H bond approaches the metal atom due to the favourable molecule conformation [45].

In such complexes the M . . . H distance is close to 1.9–2.1 Å and the interaction may be viewed as a three-centre two-electron bond including C, H, and Ni or Mo [46]. Such a bond may be called a 'soft hydrogen bond' (to differentiate it from usual 'hard hydrogen bond' between negatively charged H and electronegative atoms, such as N, O or F). More examples of such M . . . H— bonding have been found recently for complexes of titanium [47], iron [48], manganese [49], palladium [50], rhodium [51], and iridium [52].

The availability of an empty low-lying orbital at the metal is essential for all these complexes. A trend to reaching a filled electron shell is, by all evidence, the driving force of the M . . . H interaction.

References

1. R. Tang and J. K. Kochi: *J. Inorg. Nucl. Chem.*, **35**, 3845 (1973).
2. R. A. Sheldon and J. K. Kochi: *Oxidation and Combustion Reviews*, **5**, 135 (1973).
3. L. G. Makarova and A. N. Nesmeyanov: *Metody elementoorganicheskoi khimii. Rtut'* (Methods of organoelement chemistry. Mercury), p. 62, Nauka, Moscow (1965).
4. O. A. Reutov and I. P. Beletskaya: *Reaction Mechanisms of Organometallic Compounds*, p. 177, Elsevier, North-Holland Co. (1968).
5. C. Perrin and F. H. Westheimer: *J. Amer. Chem. Soc.*, **85**, 2773 (1963).
6. A. G. Kresge and J. F. Brennan: *Proc. Chem. Soc.*, 215 (1963).
7. H. C. Brown and G. Goldman: *J. Amer. Chem. Soc.*, **84**, 1650 (1962).
8. V. P. Glushkova and K. A. Kocheshkov: *Dokl. Akad. Nauk SSSR*, **116**, 233 (1957).
   V. P. Glushkova and K. A. Kocheshkov: *Izv. Akad. Nauk SSSR, ser. khim.*, 1186 (1957).
9. F. R. Preuss and I. Jaushen: *Arch. Pharm.*, **293**, 933 (1960).
   R. O. C. Norman and R. Taylor: *Electrophilic Substitution in Benzoid Compounds*, p. 200, Elsevier, Amsterdam, London, New York (1965).
   D. R. Harvey and R. O. C. Norman: *J. Chem. Soc.*, 4860 (1964).
10. M. S. Kharasch and M. S. Osbell: *J. Amer. Chem. Soc.*, **53**, 3053 (1931).
    P. W. J. de Graaf, J. Boersma, G. J. M. Van der Kerk: *J. Organomet. Chem.*, **105**, 399 (1976).

11. P. M. Maitlis: *The Organic Chemistry of Palladium*, v. 2, Academic Press, New York, London (1971).
    I. I. Moiseev: *Zhurnal Vses. Khim. Ob-va im. Mendeleeva*, **22**, 30 (1977).
    I. V. Kozhevnikov and K. I. Matveev: *Usp. khim.*, **47**, 1231 (1978).
12. R. Van Helden and G. Verberg: *Rec. Trav. Chim. Pays-Bas*, **84**, 1263 (1965).
13. I. V. Kozhevnikov, S. Ts. Sharapova, and L. I. Kurteeva: *Kinet. katal.*, **23**, 352 (1982).
14. I. I. Moiseev: *π-Kompleksy v zhidkofaznom okislenii olefinov* (π-Complexes of Olefin Oxidation in Liquid Phase), Nauka, Moscow (1970).
15. G. W. Parshall: *Homogeneous Catalysis*, J. Wiley and Sons, New York, p. 201 (1980).
16. A. K. Yatsimirskii, A. D. Ryabov, and I. V. Berezin: *J. Molec. Catal.*, **4**, 151 (1978).
17. J. L. Garnett and R. J. Hodges: *J. Amer. Chem. Soc.*, **89**, 4546 (1967).
18. G. W. Parshall: *Catalysis*, **1**, 334 (1977).
19. J. Chatt and J. M. Davidson: *J. Chem. Soc.*, 843 (1965).
20. S. D. Ittel, C. A. Tolman, A. D. English, and J. P. Jesson: *J. Amer. Chem. Soc.*, **98**, 6073 (1976).
21. a. C. Gianotti and M. L. H. Green: *J. Chem. Soc., Chem. Commun.*, 1114 (1972);
    b. M. L. H. Green and P. J. Knowles: *J. Chem. Soc.*, A, 1508 (1971).
22. a. J. M. Kliegman and A. C. Cope: *J. Organomet. Chem.*, **16**, 309 (1969).
    b. J. F. Van Baar, K. Vrieze, and D. J. Stufkens: *J. Organomet. Chem.*, **85**, 249 (1975);
    c. S. Komiya, T. Ito, M. Cowie, A. Yamamoto, and J. A. Ibers: *J. Amer. Chem. Soc.*, **98**, 3874 (1976).
23. G. W. Parshall: *Acc. Chem. Res.*, **8**, 113 (1975).
24. E. A. Grigorjan, F. S. Dyachkovskii, S. Ya. Zhuk, and L. I. Vyshinskaya: *Kinet. katal.*, **19**, 1063 (1978).
25. E. I. Heiba, R. M. Dessau, and W. J. Koehl, Jr.: *J. Amer. Chem. Soc.*, **91**, 6830 (1969).
26. M. K. Starchevskii, M. N. Vargaftik, and I. I. Moiseev: *Kinet. katal.*, **20**, 1163 (1979);
    idem, ibid, **21**, 1451 (1980);
    idem, ibid, **22**, 622 (1981).
27. K. Elmitt, M. L. H. Green, R. A. Forder, I. Jefferson, and K. Prout: *J. Chem. Soc., Chem. Commun.*, 747 (1974).
28. R. Stewart and U. A. Spitzer: *Canad. J. Chem.*, **56**, 1273 (1978).
29. B. K. Teo, A. P. Ginsberg, and J. C. Calabrese: *J. Amer. Chem. Soc.*, **98**, 3027 (1976).
30. L. Farrugia and M. L. H. Green: *J. Chem. Soc., Chem. Commun.*, 416 (1975).
31. M. L. H. Green, M. Berry, C. Couldwell, and K. Prout: *Nouv. J. Chim.*, **1**, 187 (1977).
32. W. Beck, K. Schorpp, and F. Kern: *Angew. Chem. Int. Ed.*, **10**, 66 (1971).
33. G. W. Parshall: *Acc. Chem. Res.*, **3**, 139 (1970).
34. A. J. Carty: *Organomet. Chem. Rev.*, A, **7**, 191 (1972).
35. M. I. Bruce: *Angew. Chem.*, **89**, 75 (1977).
36. A. K. Yatsimirskii: *Zhurnal neorganich. khim.*, **24**, 2711 (1979).
37. A. J. Cheney, B. E. Mann, B. L. Shaw, and R. M. Slade: *J. Chem. Soc.*, A, 3833 (1971).
38. A. J. Cheney and B. L. Shaw: *J. Chem. Soc., Dalton Trans.*, 754 (1972); *idem, ibid*, 860 (1972).
39. A. A. Kiffen, C. Masters, and L. Raynan: *J. Chem. Soc., Dalton*, 853 (1975).
40. P. A. Kramer and C. Masters: *J. Chem. Soc., Dalton Trans.*, 849 (1975).
41. F. A. Cotton, B. A. Frenz, and D. L. Hunter: *J. Chem. Soc., Chem. Commun.*, 755 (1974).
42. J. W. Rathke and E. L. Muetterties: *J. Amer. Chem. Soc.*, **97**, 3272 (1975).
43. D. M. Roe, P. M. Bailey, K. Moseley, and P. M. Maitlis: *J. Chem. Soc., Chem. Commun.*, 1273 (1972).

44. G. J. Olthof and F. Van Bolhuis: *J. Organomet. Chem.*, **122**, 47 (1976).
45. S. Trofimenko: *J. Amer. Chem. Soc.*, **89**, 6288 (1967); *idem, ibid*, **90**, 4754 (1968); S. Trofimenko: *Inorg. Chem.*, **9**, 2493 (1970).
46. F. A. Cotton, T. LaCour, and A. G. Stanislowski: *J. Amer. Chem. Soc.*, **96**, 754 (1974).
47. Z. Dawood, M. L. H. Green, V. S. B. Mtetwa, and K. Prout: *J. Chem. Soc., Chem. Commun.*, 802 (1982).
48. G. M. Dawkins, M. Green, A. G. Orpen, and F. G. A. Stone: *J. Chem. Soc., Chem. Commun.*, 41 (1982).
49. M. Brookhart, W. Lamanna, and M. B. Humphrey: *J. Amer. Chem. Soc.*, **104**, 2117 (1982).
50. T. Hosokawa, T. Ohta, S.-I. Murahashi, and M. Kido: *J. Organomet. Chem.*, **228**, C55 (1982).
51. C. A. Johnson and A. J. Nielson: *Polyhedron*, **1**, 501 (1982).
52. O. W. Howarth, C. H. McAteer, P. Moore, and G. E. Morris: *J. Chem. Soc., Chem. Commun.*, 506 (1981).

CHAPTER II

REACTIONS OF ALKANES WITH COMPOUNDS OTHER THAN METAL COMPLEXES

II.1. Reactions of Alkanes with Electrophiles [1]

Alkanes are too weak as bases, and too weak as nucleophiles, to react with the majority of acids and electrophilic reagents, as compared with olefins and aromatic hydrocarbons which, under the same conditions, can be protonated and get involved in various reactions, such as electrophilic addition or substitution. However, Olah *et al.* (see reviews [1] and refs. therein) showed that in the presence of 'superacids' (such as FSO_3H-SbF_5 or HF-SbF_5) the C—H bonds of alkanes become basic enough to coordinate protons, and that the carbonium ions $C_nH_{2n+3}^+$ produced are capable of reactions such as izomerization, fragmentation, cyclization and electrophilic substitution. The electrophilic substitution in alkanes where strong electrophilic reagents are present may include protolysis, H—D isotope exchange, alkylation, nitration and halogenation. The simplest carbonium ion to be produced upon alkane protonation is the methonium ion CH_5^+ which was initially observed in the gas phase in a secondary process in the ionization source of a mass-spectrometer [2].

$$CH_4^+ + CH_4 \longrightarrow CH_5^+ + CH_3$$

and later was proposed as an intermediate in methane interactions with solutions of superacids.

According to theoretical calculations the favoured structure of CH_5^+ corresponds to C_s symmetry (front-side protonated form)

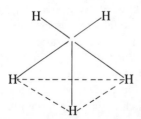

Protonation of methane via CH_5^+ leads to its protolysis

$$CH_4 + H^+ \rightleftharpoons \left[CH_3 \cdots \begin{matrix} \cdot H \\ \cdot \cdot H \end{matrix} \right]^+ \rightleftharpoons CH_3^+ + H_2$$

and, the second stage being reversible, this mechanism accounts for H–D exchange between CH_4 and D_2 in the presence of superacids.

(Dotted lines here and below symbolize the bonding orbitals of the three-centered bonds and their point of junction does not represent an additional atom).

An analogous scheme has been proposed to account for exchange of dihydrogen (or dideuterium) with a solvent medium. In this case, the hydrogen molecule becomes a base

$$\begin{matrix} H \\ | \\ H \end{matrix} + D^+ \rightleftharpoons \left[\begin{matrix} H \cdot \\ \\ H \cdot \end{matrix} \cdots D \right]^+ \rightleftharpoons \begin{matrix} H \\ | \\ D \end{matrix} + H^+$$

Reactions between alkanes and electrophilic reagents involve both the C—H and C—C bonds according to the scheme:

$$R_3H + E^+ \rightleftharpoons \left[R_3C \cdots \begin{matrix} \cdot E \\ \cdot \cdot H \end{matrix} \right]^+ \begin{matrix} \rightleftharpoons R_3C^+ + EH \\ \rightleftharpoons R_3CE + H^+ \end{matrix}$$

$$R_3C—CR_3 + E^+ \rightleftharpoons \left[\begin{matrix} R_3C \cdot \cdot \cdot \cdot CR_3 \\ \vdots \\ E \end{matrix} \right]^+ \rightleftharpoons R_3CE + R_3C$$

where $E^+ = D^+, H^+, R^+, NO_2^+, Hal^+$.

Electrophilic substitution proceeds along the highest electron density in σ-C—H or σ-C—C bonds via frontal attack. Unlike nucleophilic substitution (in which the nucleophile approaches from the opposite side of the atom being substituted) and the radical reaction (in which the activated complex appears to have a linear structure) the electrophile attacks the alkane molecule at an angle to the bond being broken, with retention of configuration of the reacting molecules. This has been confirmed, for instance, by proton substitution in adamantane where inversion of configuration is impossible.

Normally, the steric influences are not too strong, which follows from the significant reactivity of the C—C bond which is sterically more hindered than the C—H bond. The reactivity of the C—H and C—C bonds in branched alkanes increases in the sequence

$$\textit{tert-C—H} > \textit{C—C} > \textit{sec-C—H} \gg \textit{prim-C—H}$$

which corresponds to the weakening of the donor properties of the C—H bond from *tert*-C—H to *prim*-C—H and to the decrease in the C—H bond energy.

In reactions with neopentane, both C—H and C—C bonds become split.

At 80°C in the FSO_3H-SbF_5 solution, cleavage of the C—H bond takes place, while, at higher temperatures and changing to $HF-SbF_5$, preferential cleavage of the C—C bond occurs. The differences in the systems appear to depend on the size of the active species: $H_2SO_3F^+$ in FSO_3H-SbF_5, and H_2F^+ in $HF-SbF_5$. Consequently, steric effects become more pronounced as the size of the group bound to proton in the active species increases.

The attack at a *tert*-C—H bond is sterically hindered if a quaternary carbon atom is spaced too close to this bond, and the following sequence emerges:

$$Me_3C\!-\!H > Me_3C\!-\!\underset{\underset{\displaystyle Me}{|}}{\overset{\overset{\displaystyle Me}{|}}{C}}\!-\!H > (Me_3C)_2C\!-\!H \gg (Me_3C)_3C\!-\!H$$

The *tert*-C—H bond in the last compound seems to be totally unreactive so that reactions will only proceed by the C—C and primary C—H bonds.

Carbenium ions ($C_nH_{2n+1}^+$) which are formed in solutions of olefins in super-acids also interact with both the C—H and C—C bonds to bring about alkylation (substitution of H^+) or 'alkylolysis' (substitution of R^+) [3]

$$CH_3^+ + CH_4 \rightleftharpoons \left[H-\overset{\overset{\displaystyle H}{|}}{\underset{\underset{\displaystyle H}{|}}{C}} \cdots \begin{matrix} \cdot H \\ \cdot CH_3 \end{matrix} \right] \overset{-H^+}{\rightleftharpoons} CH_3CH_3$$

$$R_3'CH + {}^+CR_3 \rightleftharpoons \left[R_3'C \cdots \begin{matrix} \cdot H \\ \cdot CR_3 \end{matrix} \right]^+ \begin{matrix} \nearrow R_3'CCR_3 + H^+ \\ \searrow R_3'C^+ + CR_3H \end{matrix}$$

Nitration of alkanes under the action of nitronium salts in aprotic media (CH_2Cl_2-sulfolane) yields aliphatic nitrocompounds (Table II.1) [4].

TABLE II.1
Nitration and nitrolysis of alkanes with $NO_2^+PF_6^-$ in a CH_2Cl_2-sulfolane solution at 25°C

Hydrocarbons	Product nitroalkanes and their molar ratios	
Methane	CH_3NO_2	
Ethane	$CH_3NO_2 > C_2H_5NO_2$,	2.9:1
Propane	$CH_3NO_2 > C_2H_5NO_2 > 2\text{-}(NO_2)C_3H_7 > 1\text{-}(NO_2)C_3H_7$	2.8:1:0.5:0.1
iso-Butane	tert-$(NO_2)C_4H_9 > CH_3NO_2$	
n-Butane	$CH_3NO_2 > C_2H_5NO_2 > 2\text{-}(NO_2)C_4H_9 > 1\text{-}(NO_2)C_4H_9$	5:4:1.5:1
Neopentane	$CH_3NO_2 > $ tert-$C_4H_9NO_2$	3.3:1
Cyclohexane	nitrocyclohexane	
Adamantane	1-nitroadamantane > 2-nitroadamantane	17.5:1

The alkane σ-basicity is apparently lower than the n-basicity of donor molecules used as solvents. Nitroalkanes are also n-donors and, hence, slow down the nitration.

All this results in low product yields: 0.1% for methane, 2–5% for higher alkanes and isoalkanes and 10% for adamantane at 25°C [4]. Alkane chlorination at −78°C in the presence of strong acids (for instance, in the system Cl_2–SbF_6–SO_2ClF) produces chlorosubstituted derivatives; notably, with methane, only methyl chloride is produced, with neither CH_2Cl_2 nor $CHCl_3$ being observed. The implication is that, in electrophilic chlorination by a Cl^+ ion, the electronegative Cl atom introduced into the methane molecule deactivates it and makes it incapable of further chlorination.

The electrophilic oxygenation of alkanes with ozone and hydrogen peroxide takes place readily in the presence of superacid media [1b]. The reaction proceeds via an electrophilic attack by ions HO_3^+ and $H_3O_2^+$, respectively, through pentacoordinated carbonium ions, e.g.,

$$CH_3-\underset{\underset{CH_3}{|}}{\overset{\overset{CH_3}{|}}{C}}-H \xrightarrow{H_3O_2^+} \left[CH_3-\underset{\underset{CH_3}{|}}{\overset{\overset{CH_3}{|}}{C}} \cdots \overset{\cdot \cdot H}{\underset{\cdot \cdot OH}{}} \right]^+ \xrightarrow{-H^+} CH_3-\underset{\underset{CH_3}{|}}{\overset{\overset{CH_3}{|}}{C}}-OH$$

$$H^+ \Big| -H_2O$$

$$\underset{CH_3}{\overset{CH_3}{}}\!\!\diagdown\!\!\underset{}{\overset{}{C}}\!\!=\!\!O\!\!\diagup\!\!\overset{CH_3}{} \xleftarrow{-H_2O} CH_3-\underset{\underset{CH_3}{|}\,\underset{H}{|}}{\overset{\overset{CH_3}{|}}{C}}-\overset{+}{O}-OH \xleftarrow{H_2O_2} CH_3-\underset{\underset{CH_3}{|}}{\overset{\overset{CH_3}{|}}{C^+}}$$

Strong acids induce alkane isomerization. The primary step here involves protolysis of the C—H or C—C bond to form a carbonium ion. At a later stage dihydrogen, or an alkane molecule with low molecular weight, splits off to produce an alkyl-carbenium ion which can subsequently participate in intra- or intermolecular processes.

The isomerization of butane into *iso*-butane proceeds via formation of protonated cyclopropane as an intermediate

$$Me-CH_2-CH_2-Me \underset{}{\overset{+H^+}{\rightleftharpoons}} \left[Me-CH_2-\underset{\overset{\cdot\cdot}{\overset{Y}{H\ \ H}}}{CH}-Me \right]^+ \underset{+H_2}{\overset{-H_2}{\rightleftharpoons}}$$

$$MeCH_2\overset{+}{C}HMe \rightleftharpoons \left[\underset{CH_2-CH-Me}{\overset{CH_3}{\overset{\cdot\cdot}{\diagup\diagdown}}} \right]^+ \rightleftharpoons$$

$$\overset{+}{C}H_2-\underset{}{\overset{\overset{CH_3}{|}}{CH}}-Me \rightleftharpoons Me_3C^+ \underset{+H^+,\ -H_2}{\overset{+H_2,\ -H^+}{\rightleftharpoons}} Me_3CH$$

Intermolecular alkylation processes are started by isomerizations which pass through an expansion or contraction of the ring, as for instance, does the cyclo-hexane to methylcyclopentane isomerization.

II.2. Reactions of Alkanes with Atoms and Free Radicals

Atoms and radicals, being short-lived active intermediates in various chemical reactions, react readily with various molecules including alkanes. Reactions of alkanes with atoms and radicals take part in many radical-chain processes in gas phase and solutions, e.g., cracking, halogenation, oxidation, etc.

A most typical reaction between alkanes or other C—H bond-containing molecules and atoms or radicals is the abstraction of an H atom

$$R \cdot + R'H \longrightarrow RH + R' \cdot$$

Normally, the activation energies in such elementary reactions are low, not exceeding $10-15$ kcal mole^{-1}. For the same radical, the activation energy depends linearly on the bond dissociation energy (for the particular case of H abstraction from alkanes this will be the C—H bond dissociation energy) corresponding to the Evance—Polanyi rule:

$$\Delta E = \alpha \Delta D$$

Thus, for instance, branched hydrocarbons would show a 'normal' selectivity of C—H bond attack: $3° > 2° > 1°$. The selectivity controlling factor α depends on the atom or radical nature and varies within the range $0-1$.

For many radicals, α approaches 0.25 so that, in a number of similar reactions, the relationship

$$E = A - 0.25 \, q$$

is applicable, where q is the reaction enthalpy, A is a constant approaching 11.5 (the Polanyi—Semenov rule) [5]. In reactions where H is abstracted from alkanes and their derivatives with polar substituents, radicals behave as moderately strong electrophiles.

Analysing the role that the polar substituents play in reactions with radicals, Shteinman [6] concluded that the effect of substituents can be described by the Taft two-parameter correlation equation:

$$\log k/k_0 = \rho^* \sigma^* + n\psi$$

where ψ is the factor of the substituent conjugation with the reaction centre, n is the number of substituents capable of such substitutions. Table II.2 lists the values of ρ^* and ψ for several atoms and radicals. The physical meaning of ψ corresponds to the stabilization of the free radicals $R \cdot'$ produced, provided that the substituent becomes conjugated with the unpaired electron. The steric

TABLE II.2

Values of ρ^* and ψ for atoms and radicals in the
reactions $R\cdot + R'H \longrightarrow RH + R'\cdot$

R	ρ^*	ψ
Cl	−1.8	4.0
Br	−2.7	7.0
I	−3.4	8.0
H	−2.6	5.0
O	−2.0	4.5
OH	−1.0	2.2
CF_3	−1.4	2.9
CCl_3	−2.5	6.2
CH_3	−1.5	3.2

factors in reactions with simple radicals seem to be insignificant, as is shown by the $\Delta E = \alpha \Delta D$ relation and the absence of the relevant term from the correlation equation. The negative values of ρ^* in Table II.2 suggest, as has already been noted, that the action of free radicals on the C—H bond is electrophilic in character. However, absolute values of ρ^* are not high so that the introduction of electronegative substituents would not considerably decrease the reaction rate constant during the H atom abstraction (actually, the constant can even increase owing to the conjugation effect).

Polar factors may become far more important in cation-radical reactions in which the unpaired electron is supplemented by a positive charge. Aminium radicals are a good example in this regard.

The simplest aminium ion-radicals NH_3^+ can be obtained in the reaction of Ti^{3+} ions with hydroxylamine in acidic media [7, 8]

$$Ti^{3+} + NH_2OH \xrightarrow{H^+} NH_2^+ + Ti^{4+} + H_2O$$

The pK_a value for NH_3^+ is 6.7 [8] which implies that aminoradicals NH_2^+ should be protonated to form aminium ion-radicals in sufficiently acidic media. These radicals react readily with compounds containing C—H bonds to abstract the H atom. Thus, the reactions between simple aminium ion-radicals and alcohols, such as

$$NH_3^+ + CH_3OH \longrightarrow NH_4^+ + \cdot CH_2OH$$

are fast, although the selectivity is higher than with more active $\cdot OH$ radicals: in the NH_3^+-involving reactions only α-radicals have been observed (for instance, $CH_3\dot{C}HOH$ in a reaction with ethanol), whereas $\cdot OH$-involving reactions produce both α- and β-radicals.

A greater selectivity of attack on the C—H bond has been noted for sterically hindered radicals $R_3 N^{\cdot+}$ containing organic groups R. They can be formed through a reaction between amino-oxides and bivalent iron [9]

$$R_3 NO + 2H^+ + Fe(II) \longrightarrow R_3 N^{\cdot+} + Fe(III) + H_2 O$$

The resulting aminium radicals can also interact with the C—H bond, with H-atom abstraction. Here the reaction can be intramolecular as, for instance, with $C_4 H_9 N(CH_3)_2 O$:

In the presence of alkanes their hydroxylation or chlorination can occur (where starting materials are N-chloramines).

A very characteristic feature of reactions which involve aminium ion-radicals is their extreme sensitivity to steric and polar factors. Thus, pentane chlorination by N-chloro-2,2,6,6-tetramethylpiperidine produces chloropentanes in which 2-chloropentane predominates: relative yields of 1-, 2- and 3-chloropentane are 9, 87 and 4% respectively. As will be shown later, the preferential interaction with the C—H bonds of a terminal (ω) and the penultimate ($\omega - 1$) hydrocarbon atom occurs mainly in the reactions of sterically hindered reagents. Where the reagent is not active enough to react with the CH_3-group having the maximum C—H bond energy, the reaction will preferentially involve the ($\omega - 1$) carbon atom. Similarly, the tert-butyl group is far less reactive towards such species (caused by steric effects due to two neighbouring methyl groups) than the terminal methyl group with no branching nearby. Thus, chlorination of 2,2-dimethylbutane and 2,2-dimethylpentane by N-chloro-2,2,6,6-tetramethylpiperidine gives a ratio of about 10—20 : 1 for the selectivity of attack of

ω-methyl chlorination in comparison with chlorination of the first carbon atom. The introduction of electronegative groups into a molecule, such as CH_2OH, $CHOHCH_3$, OR, CO_2H, etc. makes the neighbouring C—H bond totally inactive towards aminium radicals. Alcohols formed in hydroxylation by aminium radicals do not suffer further attack.

If a molecule with a long hydrocarbon chain contains an electronegative group, the reaction will take place between the aminium radicals and the C—H bond most distant from this group. For instance, 1-octanol gets chlorinated or hydroxylated mainly into position 7 (with 92% and 70% selectivity, respectively).

Cyclohexanol is chlorinated at the C-4 atom, cyclododecanone is attacked preferentially at positions C-6 and C-7, cyclopentadecanone at the C-8 atom. These features are characteristic of the biological oxidation of alkanes (Section IV.2.4). Proceeding from this, the authors of [9] consider the reactions between aminium radicals and C—H bonds to be analogous to enzymatic oxidation.

II.3. Reactions between Alkanes and Carbenes [10]

Simple carbenes are usually very reactive and, like free radicals, normally have only a transient existence. Unlike free radicals, however, carbenes can be produced and can enter into reactions in two states, i.e. singlet and triplet, corresponding to the antiparallel and parallel spins of the two electrons. One of these compounds is in the ground-state, the other is electronically excited; for the simplest carbene, methylene, the ground state is a triplet, while a singlet methylene corresponds to the excited state, the energy difference between the two states being about 19.6 kcal mole^{-1}, according to recent data [11], in contrast to a number of earlier results, which tended to give much smaller values (6—9 kcal mole^{-1}).

In photo- and thermal decomposition of singlet molecules, such as ketene or diazomethane

$$CH_2{=}C{=}O \xrightarrow[2800 \text{ Å}]{h\nu} CH_2 + CO$$

$$CH_2NN \xrightarrow{\Delta,\ h\nu} CH_2 + N_2$$

the singlet methylene is usually produced primarily, according to the spin conservation rule, and this can transform subsequently into a triplet carbene as a result of collisions with the particles present. The singlet methylene shows very high chemical activity and reacts readily with unsaturated compounds, such as olefins, acetylenes, carbon monoxide [10], and even dinitrogen [12],

$$CH_2 + CH_2{=}CH_2 \longrightarrow \overset{\displaystyle CH_2}{\overset{\displaystyle \diagup \diagdown}{CH_2{-}CH_2}}$$

$$CH_2 + CO \longrightarrow CH_2{=}C{=}O$$

$$CH_2 + N_2 \rightleftharpoons CH_2N_2$$

Methylene also reacts with molecules which contain σ-bonds only, e.g., with dihydrogen

$$CH_2 + H_2 \longrightarrow CH_4^*$$

$$CH_4^* + M \longrightarrow CH_4 + M$$

$$CH_4^* \longrightarrow CH_3 + H\cdot$$

(here M is the third body participating in the collision) and with alkanes. In the latter case, the most typical CH_2 reaction is the insertion into the C—H bond, which proceeds extremely readily with practically no activation energy. Methane reacts with methylene in the gas phase to produce ethane

$$CH_2 + CH_4 \longrightarrow C_2H_6$$

In methylene-hydrocarbon reactions taking place in the liquid phase, CH_2 attacks various C—H bonds of alkanes in a virtually nonselective way. Thus, the reaction of CH_2 with pentane produces n-hexane, 2-methylpentane and 3-methylpentane in the proportion $48:35:17$ at $-75°C$, and $49:34:17$ at $15°C$. At both temperatures this ratio closely approaches the statistical one (obtained with a number of respective C—H bonds taken into account): $50:33.3:16.7$. Reactions between singlet methylene and alkanes in the gas phase are somewhat more selective, with the 'normal' sequence prevailing: $3° > 2° > 1°$. In the reaction of CH_2 with butane and isobutane the methylene has been found to react faster with secondary and tertiary C—H bonds by 15–20% and 50%, respectively, than with the primary C—H bonds.

In reactions with amines, ethers, alkylhalogenides, the activity of the C—H bond of the α-carbon atom is high enough to demonstrate that conjugation contributes to the stability of transition state. The isotope effect in the insertion of methylene into the C—H bond is rather small. For instance, in the reaction with propane and 2,2-D_2-propane $k_H/k_D = 1.3$, in cis-2-butene and cis-2-butene-d_8 reaction $k_H/k_D = 1.96$, and $k_H/k_D = 1.55$ for CH_3 and CH groups, respectively.

Insertion of CH_2 into the C—H bond proceeds with retention of configuration. Triplet methylene is usually less reactive than the singlet state and behaves more similarly to free radicals. It can abstract the hydrogen atom from the C—H bond or add to the double bonds forming primarily trimethylene biradical with a loss of stereospecificity. At the same time, it reacts readily with such scavengers of free radicals as O_2 and NO, and the reactivity of singlet CH_2 formed together with the triplet one can be studied quantitatively in the presence of free radical scavengers.

Various other carbenes are capable of insertion into C—H bonds but usually react much more selectively than CH_2. Thus phenylcarbene reacts 6—8 times faster and CHCl twenty times faster with a secondary C—H bond of n-pentane than with a primary one. Dichlorocarbene CCl_2 reacts even more selectively and inserts only into secondary and tertiary C—H bonds. Even secondary C—H bonds are not affected if a tertiary C—H bond is present.

De More and Benson [13a] have proposed that, in the methylene insertion into the hydrocarbon C—H bond, the hydrogen atom is first attacked to produce a complex similar to that which is formed in recombination of two radicals. For instance, for methylene

$$CH_2 + CH_4 \longrightarrow CH_3 \ldots CH_3 \longrightarrow C_2H_6$$

This conclusion was supported by calculations of the potential energy surface carried out by Hoffmann et al. by the extended Hückel method [13b] and by Dewar et al. by the MINDO/2 procedure [13c].

The calculations show that a hydrogen atom of methane approaches a carbon atom of methylene to a distance of 2 Å without a noticeable increase of potential energy if the free p-orbital of CH_2 is directed towards electron-populated orbital corresponding to the σ—C—H bond in methane.

II.4. Alkane Hydroxylation by Peracids

Peracids are capable of oxidizing organic compounds by transferring an oxygen atom of a peroxide group. This reaction is best studied using olefin oxidation to yield oxides (Prilezhaev reaction) [14]. The reaction is most likely to involve an electrophilic attack of the double bond by an oxygen atom

Electron-donating substituents in olefins and electron-withdrawing ones in superacids promote the reaction, confirming the electrophilic nature of the oxygen being transferred. One of the most reactive reagents in the Prilezhaev reaction is trifluoroperacetic acid which contains a strong acceptor group CF_3. This acid has been shown to be capable of hydroxylating C—H bonds in alkanes [15].

The mechanism of alkane hydroxylation appears to be similar to that of olefin epoxidation and involves an electrophilic attack of the oxygen atom:

It has been shown for the reaction between trifluoroperacetic acid and *cis-* or *trans*-1,2-dimethylcyclohexane that the molecule retains its configuration completely (100%). The electrophilic nature of this reaction is also confirmed by the normal selectivity sequence in the attack of the C—H bonds: $3° > 2° > 1°$. Hydroxylation with trifluoroperacetic acid proceeds by the oxenoid type of the reaction (the Hamilton definition [15]) in which the oxygen atom (oxene) inserts into the C—H bond with no intermediate free radicals or ions. This mechanism is evidenced, in particular, by retention of configuration. Table II.3 summarizes results for several reagents capable of C—H bond hydroxylation.

TABLE II.3
Characteristics of certain reactions of alkanes

Reaction	Relative reactivity per H atom			Retention of configuration, %
	prim	sec	tert	
Pyridine-*N*-oxide + *hv*	1	10	40	100
Pyridine-*N*-oxide + *hv*	1	15	70	> 95
Nitrobenzene + *hv*	1	20	300	0
Ozonation	1	13	110	60−70
Carbene-O_2	1	15	140	0
Trifluoroperacetic acid	1	~30	~100	100
Carbene insertion	1	1−8	1−21	100
Nitrene insertion	1	5−15	14−70	100
Oxidation of chromate	1	100	7000	70−100
H elimination by RO	1	12	44	–
H⁻ elimination	1	~10^4	~10^8	–

A mechanistic scheme to account for the action of hydroxylating agent $X = O$ upon RH hydrocarbon has been proposed in [15]:

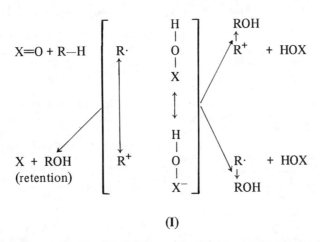

(I)

The actual pattern of hydroxylation is governed by the peculiarities of transition state or intermediate **I**. If HOX is not very stable, as a radical (HOX·) or as an ion (HOX⁻), **I** turns rapidly into alcohol ROH with retention of configuration. For relatively stable HOX, complex **I** decomposes to R· and HOX. R· may then produce ROH, which would be completely equilibrated.

According to [15] the selectivity of the reaction towards primary, secondary or tertiary bonds depends on the degree of ionic character of **I**. The selectivity of reaction **I** increases with an increase of ionic character. However, this simplified reasoning does not take into account steric factors, nor the thermodynamics of the hydroxylation reaction: even for an essentially radical (non ionic) character of **I** the selectivity must become more pronounced as the activity of the hydroxylating agent XO decreases.

References

1. a. G. A. Olah: *Angew. Chem. Int. Ed.,* **12**, 173 (1973).
   b. G. A. Olah, D. G. Parker, and N. Yoneda: *ibid,* **17**, 909 (1978).
2. V. L. Tal'roze and A. K. Lubimova: *Dokl. Akad. Nauk SSSR,* **86**, 909 (1952).
3. G. A. Olah, Y. Halpern, J. Shen, and Y. K. Mo: *J. Amer. Chem. Soc.,* **93**, 1251 (1971).
4. G. A. Olah and H. Chi-hung Lin: *J. Amer. Chem. Soc.,* **93**, 1259 (1971).
5. N. N. Semenov: *Some Problems in Chemical Kinetics and Reactivity,* Princeton Univ. Press, Princeton, N. Y., Vol. 1 (1958), Vol. 2 (1959).
6. A. A. Shteinman: *Kinet. katal.,* **15**, 831 (1974).
7. V. F. Shuvalov and A. P. Moravskii: *Kinet. katal.,* **17**, 870 (1976).
8. B. C. Gilbert and P. R. Marriot: *J. Chem. Soc., Perkin II,* 987 (1977).

9. N. C. Deno, E. J. Jedziniak, L. A. Messer, and E. S. Tomezsko: *Enzyme Action* (Bio-organic Chemistry, Vol. 1, Ed. E. E. Van Tamelen) p. 79, Academic Press, New York, San Francisco (1977).

10. P. F. Zittel, G. B. Ellison, S. V. O'Neil, E. Herbst, W. C. Lineberger, and W. P. Reinhardt: *J. Amer. Chem. Soc.*, 98, 3731 (1976).

11. W. Kirmse: *Carbene Chemistry*, Academic Press, New York, London, First Edition (1964); Second Edition (1971).

12. Yu. G. Borod'ko, A. E. Shilov, and A. A. Shteinman: *Dokl. Akad. Nauk SSSR*, 168, 581 (1966); A. E. Shilov, A. A. Shteinman, and M. B. Tjabin: *Tetrahedron Letts.*, 4177 (1968).

13. a. W. B. DeMore and S. W. Benson: *Adv. Photochem.*, 2, 219 (1964).
    b. P. C. Dobson, D. M. Hayes, and R. Hoffman: *J. Amer. Chem. Soc.*, 93 6188 (1971).
    c. N. Bodor, M. J. S. Dewar, and J. S. Wasson, *J. Amer. Chem. Soc.*, 94, 9095 (1972).

14. E. N. Prilezhaeva: *Reaktsija Prilezhaeva. Electrofil'noje okislenije* (The Prilezhaev Reaction. Electrophilic Oxidation), Nauka, Moscow (1974).

15. G. A. Hamilton, J. R. Giacin, T. M. Hellman, M. E. Snook, and J. W. Weller: *Ann. N. Y. Acad. Sci.*, 212, 4 (1973).

CHAPTER III

ACTIVATION OF ALKANES ON THE SURFACE OF METALS AND METAL OXIDES. REACTIONS OF ALKANES WITH METAL ATOMS AND IONS

At present it is becoming ever more obvious that the mechanism of heterogeneous catalytic reactions is in general analogous to the reactions in homogeneous solutions, and that metal complexes behave in a similar way on surfaces and in solution. Metal atoms or ions on surfaces can be looked upon as analogous to metal atoms or ions surrounded by certain ligands (in the particular case of metal surfaces, ligands are other metal atoms surrounding a given atom): chemisorption may be looked upon as complex formation on the surface. In those cases when a combined participation of several metal ions is essential for the process, clusters can serve as homogeneous analogs.

The advantage of metals and their simple oxides over complexes in solutions for the special case of alkanes (as well as for some other substrates) consists in the possibility of using elevated temperatures (up to several hundreds degrees) – which is essential for those processes having a comparatively high activation energy. In addition, in this case there is no such problem as C—H bonds contained in solvents or ligands which, as we have seen, can be effective competitors with alkanes in reactions where a C—H bond is broken.

All these considerations have led us to regard the known catalytic reactions of alkanes on surface in terms of looking for analogous reactions in solutions. Much work has been devoted to reactions of various hydrocarbons, including alkanes, on heterogeneous catalysts. The results of the work are summarized in a number of reviews and monographs [1–5].

Here we shall dwell upon the general characteristics of these reactions which are especially interesting from the point of view of their comparison with homogeneous activation.

The simplest heterogeneous catalytic reactions are undoubtedly those of isotope H—D exchange of alkanes with D_2, D_2O and other deuterated hydrocarbons which are catalyzed by metal oxides (Al_2O_3, Cr_2O_3, etc.) and metals (Ni, Pt, Rh, Ir, Ru, Pd, etc.). In addition, heterogeneous catalysts can catalyze such reactions as oxidation, dehydrogenation, cracking, dehydrocyclization, isomerization and others.

The last part of this chapter will deal with the results obtained recently for individual metal atom and metal ion reactions with alkanes. These may be also considered from the viewpoint of analogy with the reactions of complexes, the atom or ion being an extreme case of a complex with no ligands present at a metallic center or ion.

III.1. Isotope Exchange

Metal oxides whose surfaces are activated by heating at high temperatures catalyze H–D exchange between H_2 and D_2 and hydrogenation of olefins. The same oxides catalyze the H–D exchange of alkanes with D_2.

Thus, upon heating Cr_2O_3 *in vacuo* or under dry inert gas at temperatures over 300°C (the best results being obtained at 470°C) its surface is dehydrated and acquires the characteristics of a hydrogenation catalyst. Chromium oxide is also a catalyst of H–D exchange between alkanes and D_2 which occurs readily at temperatures higher than 200°C.

Cycloalkanes exchange more readily than linear alkanes. For example, cyclopropane exchanges even at room temperature. Molecules containing one D atom prevail in the initial stages of the reaction. Benzene is also exchanged with D_2, the reaction proceeding readily at 80°C. Here, d_1 molecules are also formed initially. Active centers which catalyze the exchange of various types of hydrocarbons and other reactions on the surface of Cr_2O_3 can be different. For example, the H–D exchange of benzene and hydrogenation of cyclohexene do not inhibit each other, hence these reactions might be thought to be catalyzed by different centers.

When toluene reacts with D_2 on Cr_2O_3, the hydrogen atoms of the ring are exchanged about one order of magnitude faster than those of the methyl group. At 80°C for benzene, methyl hydrogen of toluene, cyclopropane and cyclohexane the following order of exchange rates is observed: 1 : 0.1 : 0.02 : 0.0002 [4].

For alkanes, the H–D exchange with D_2 was observed to proceed somewhat faster for primary than for secondary C–H bonds [2]. This can be explained by a carbanion character of the intermediate species formed on the oxide surface

$$\begin{array}{ccc}
\diagdown\!\!\diagup & & \diagdown\!\!\diagup \\
\text{C} & & \text{C}^{\circ} \\
| & & \vdots \\
\text{H} & & \vdots \\
& & \vdots \\
Cr^{3+}O^{2-} & \rightleftharpoons & Cr^{3+}OH^{-}
\end{array}$$

Carbanions with negative charge on the primary C atom are known to be the most stable (see Introduction). The possibility of steric hindrance should be also taken into account, being of minimum significance for primary C—H bonds.

After heating *in vacuo* at 450–700°C, the aluminium oxide becomes capable of catalyzing the H–D exchange between alkane (C_3H_8, C_2H_6) and D_2 and between alkanes themselves (e.g., CH_4 and CD_4). The activation energy of CH_4/CD_4 exchange is as low as 5.7 kcal mole^{-1}. The activity of the catalyst is selectively poisoned by addition of CO_2, NO, C_2H_4, C_3H_6, whereas in olefin isomerization CO_2 does not inhibit the reaction.

The dehydrated aluminium oxide can also catalyze the isotope exchange of olefins (C_2H_4/C_2D_4) and aromatics (C_6H_6/C_6D_6), as well as C_6H_6 and D_2, the reactivity of C—H bonds decreasing in the order: aromatic ring > olefin vinyl group > alkanes.

The activation energies of propane and benzene exchange with D_2 are 8.7 and 6.0 kcal mole^{-1}, respectively, on aluminium oxide prepared from hibbsite; for methane and benzene exchange on Al_2O_3 from aluminium isopropoxide they are equal to 5.7 and 4.3 kcal mole^{-1}, respectively [4].

Usually in reactions of hydrocarbon H–D exchange with dideuterium SiO_2–Al_2O_3 is less active than Al_2O_3, and the activity in comparison with pure Al_2O_3 is particularly low for alkanes. For 13% Al_2O_3–SiO_2 the activation energies of methane and benzene are 33 and 10 kcal mole^{-1}, respectively [4].

The other oxides (Co_3O_4, NiO, ZnO, Fe_2O_3) can also catalyze various types of isotope exchange.

The isotope H–D exchange of hydrocarbons with D_2 on metal surfaces was found soon after the discovery of deuterium in the first works on isotope exchange by Farcas and Taylor, Morikava, Horiuti, and Polanyi. Metal films and wires were used as catalysts. The isotope exchange occurs even at room temperature, though it is generally studied under heating conditions. As early as in the first papers the isotope exchange was stated to proceed more readily than cracking, though the C—C bond was markedly less stable than the C—H one.

Aromatic and alkylaromatic hydrocarbons exchange with D_2 more readily than alkanes. The exchange of hydrogen atoms of C—H bonds of the aromatic ring is particularly easy, except at the *ortho*-position to the substituent, where some steric hindrance evidently arises. A low reactivity in the exchange with D_2 or C—H in the *ortho*-position to the substituent was observed, for example, for supported nickel films at temperatures about 0°C and in the exchange with D_2O for metallic platinum. A higher reactivity in the side chain of the benzene ring is shown by the C—H bond in the α-position to the benzene ring.

The most active among the alkanes are cycloalkanes, C_5H_{10} being usually somewhat more active than C_6H_{12}. The open-chain alkanes, starting from C_2H_6 onwards, have similar reactivities, higher hydrocarbons being slightly more reactive than ethane. Neopentane and particularly methane are the least reactive.

For the H–D exchange of propane with D_2 the following order of metal reactivities was found:

$$Pt > Rh > Ir > Ru > Pd$$

which coincides with that found for hydrocracking. However, depending on the method of preparation, additions and reaction conditions, catalyst activity and order of reactivity for various metals may vary.

The mechanism of a surface H–D exchange reaction must generally involve the cleavage of the C–H bond (dissociative mechanism) and the formation of an alkyl group bound to the surface metal atom, C–H cleavage being probably the most difficult stage. The dissociation of the hydrogen molecule normally occurs much more easily than that of hydrocarbons. This assertion is supported by the isotope effect (the usual rate of exchange between RH and D_2 is markedly higher than between RD and H_2), and the largest isotope effect among hydrocarbons is usually observed among the alkanes.

A very important feature of the heterogeneous isotope exchange of hydrocarbons with D_2 or D_2O on metal surfaces is the ability of many catalysts to carry out the exchange of several hydrogen atoms in a hydrocarbon molecule within one contact with the catalyst center (a so-called multiple exchange). This type of exchange is characterized by a multiple exchange factor (M) denoting the average number of H atoms exchanged within one contact. To differentiate between multiple exchange and isotope exchange where $M = 1$ the latter is called stepwise exchange.

The study of the relation of multiple exchange to molecular structure shows that it proceeds most readily in the presence of β-C–H bonds, which can be explained by β-elimination leading to an olefin-hydrido complex either with a single metal atom

or with the participation of a neighbouring M atom

If hydride H can exchange with D and the rate of β-elimination is comparable to or higher than that of hydrocarbon desorption, then multiple exchange will occur. The importance of intermediate complex formation with an olefin is confirmed by the absence of exchange via a quaternary carbon atom.

Thus, in the exchange of C_3H_8 mixtures with D_2 (1 : 6) on Pd at 146°C, C_3D_8 (50%), C_3D_7H (12%), and $C_3D_6H_2$ (4%) are formed initially. At the same time, neopentane (I), 3,3-dimethylpentane (II) and 2,2,3-trimethylpentane (III) give the following molecules with the maximum deuterium number for each of these molecules [2, 4]

$$(CH_3)_3CCH_2D \qquad C_2H_5C(CH_3)_2C_2D_5 \qquad (CH_3)_3CCD(CD_3)_2$$

$$\textbf{(I)} \qquad\qquad \textbf{(II)} \qquad\qquad \textbf{(III)}$$

Cyclopentane in the isotope exchange with D_2 catalyzed by metals of the Group VIII elements provides a large amount of D_5 molecules with low D_2 content. That means that hydrogen atoms of one side of the ring are primarily exchanged in the chemisorption of this molecule. For various metals the factor M may vary greatly. For example, only small quantities of heavily exchanged cycloalkanes are produced on rhodium whereas, in the exchange on palladium, a great amount of cycloalkanes with strong multiple exchange is formed. Here along with the maximum on cyclopentane, D_5, the second maximum for completely exchanged cyclopentane, D_{10}, is also observed [1]. When increasing the temperature, this second maximum increases and becomes prevalent at temperatures higher than 100°C. Thus, a mechanism has been postulated according to which the cyclopentane molecule turns over, without losing the link with the surface catalyst atoms. This mechanism can involve the so-called 'rolling over' process where the olefin is intermediately bound to the four surface centers(*).

The second mechanism proposed involves an intermediate π-allylic complex formation. In the 'rolling over' mechanism, as is clear from the scheme, two neighbouring H atoms must remain unexchanged, which is actually the case for some catalysts (the maximum at D_8 is observed on palladium [6] and platinum [7]). At the same time, the maximum on Ni and Ni–Cu catalysts is observed at D_{10}, indicating that the π-allylic mechanism is probably operating here [8].

The formation of π-bound olefin is not the only path leading to the multiple exchange of hydrocarbons, as is clear in view of methane multiple exchange, where such a path is impossible. The example of distribution of methane isotopic molecules formed in the H–D exchange of CH_4 with D_2 on rhodium at $162°C$ is given in Fig. III.1 [4].

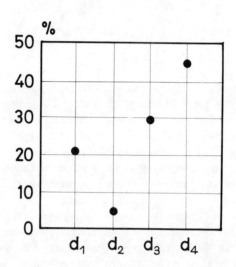

Fig. III.1: The distribution of products in the exchange of CH_4/D_2 on Rh at $162°C$ (calculated [4] from the results of [9]).

The stepwise exchange (resulting in methane-D_1) and multiple exchange of D_1–D_4 have been shown to proceed via different mechanisms, supported by some kinetic data. Methane exchange on transition metals (Ni, Pd, Pt, Rh, W) is characterized by negative order with respect to molecular deuterium, the multiple exchange being more strongly inhibited by D_2 than the stepwise one [9].

It is assumed that the subsequent dissociation of the surface methyl group to form a chemisorbed carbene is necessary for the stepwise exchange to take place. The activation energy for the multiple exchange is higher than for the stepwise one, i.e. CH_3 dissociation evidently requires greater energy than does CH_3 formation from CH_4. Upon the formation of adsorbed carbene $(CH_2)_{ads}$, a rapid exchange to produce $(CD_2)_{ads}$ occurs leading to the appearance of $(CD_3)_{ads}$ and free CD_4, which explains the maximum CD_4 formation.

III.2. Dehydrogenation and Dehydrocyclization

The mechanism of alkane isotope exchange is closely connected with that of catalytic dehydrogenation and dehydrocyclization.

The formation of alkyl groups from alkanes on metal surfaces leading to the formation of absorbed olefinic molecules may, on their desorption at sufficiently high temperatures, finally lead to the reaction of dehydrogenation of alkanes. The latter is experimentally observed on a number of metals and oxides in the range of 350–650°C and usually proceeds in the presence of dihydrogen.

Catalytic dehydrogenation is an important industrial method of hydrocarbon processing (e.g., butene and butadiene production from butane), in view of which a great number of papers has been devoted to catalytic dehydrogenation (see, e.g., reviews [1, 10, 11]).

Among metals, the highest catalytic activity is shown by platinum metals, such as Pt, Pd, Rh, Ir.

The reaction of dehydrogenation is often followed by hydrogenolysis and cracking (e.g., demethanation).

From the point of view of the reaction mechanism, an interesting result was obtained when studying propane dehydrogenation to propylene on an alloy of gold and platinum at temperatures of 360–390°C with platinum concentration in the alloy equal to 0.5–14.0 at % [12]. The specific rate in this temperature region was shown to change proportionally with the platinum content in the alloy. This result leads to the important conclusion that only one platinum atom takes part in the rate determining step. The authors of [12] consider their data to be in a very good agreement with the following mechanism

Stage 2 is a rate-determining step.

It should be mentioned that, in accordance with this scheme, the mechanism for the reverse process of catalytic olefin hydrogenation is analogous to the currently accepted mechanism of catalytic olefin hydrogenation in homogeneous solutions with the participation of mononuclear platinum complexes.

Heterogeneous catalytic dehydrocyclization of alkanes of sufficiently long chain length occurs together with dehydrogenation. For example, the following reactions take place with n-heptane on a catalyst containing 0.6% Pt on γ-Al$_2$O$_3$

The catalytic C$_6$-dehydrocyclization of alkanes evidently involves a number of stages

$$\text{alkane} \longrightarrow \text{olefin} \longrightarrow \text{diene} \longrightarrow \text{triene} \longrightarrow \text{alkylcyclo-}$$
$$\text{hexadiene} \longrightarrow \text{aromatic hydrocarbon}$$

C$_5$-Dehydrocyclization proceeds both on platinum and aluminochromium catalysts [3].

The study of *n*-heptane (^{13}C-1) dehydrocyclization catalyzed by Pt–Al$_2$O$_3$ by observing the ^{13}C distribution in the products formed demonstrates that toluene formation proceeds via a mechanism of direct 1–6 cyclization or by a more complicated mechanism of ring formation and 1–5 ring opening followed by 1–6 cyclization [13].

The study of the catalytic reactions of isomerization, dehydrocyclization, and hydrogenolysis of *n*-pentane and *n*-hexane on alloys of gold and platinum shows that isomerization (as well as dehydrogenation) proceeds on separate platinum atoms, whereas for dehydrocyclization and, particularly, hydrogenolysis the presence of several platinum atoms is required in the active center [14]. It should be noted that in many cases the interpretation of the data obtained is complicated due to some uncertainties typical of heterogeneous catalysis, such as the role of admixtures covering the catalyst surface, the differing nature of active centers on the surface, etc.

Modern surface diagnostic techniques can be used to determine the nature of catalytically active centers on the surface of a catalyst [15]. Auger and X-ray photoelectron spectroscopy were used to determine the surface composition and the oxidation state of atoms in the active catalyst surface, and low-energy diffraction to determine the atomic surface structure. These investigations have shown that the active catalyst must have several active sites distinguishable by the number of their nearest neighbours or by their oxidation state. Atomic steps and kinks (surface irregularities) on various transition metal surfaces are active in breaking C–H, C–C and H–H bonds. The active surface was found to be largely covered with a carbonaceous deposit, which can strongly influence the catalyst activity.

III.3. Heterogeneous Oxidation [16]

The catalytic oxidation of alkanes on heterogeneous catalysts is potentially a very important field, but even now it is still at the beginning of its development, particularly in comparison with the oxidation of olefins and aromatic hydrocarbons, which is already widely used in chemical industry. At present there are a few industrial alkane oxidation processes, e.g., the synthesis of maleic anhydride by oxidation of butane, and the synthesis of 1,3-butadiene by oxidative dehydrogenation of butane.

The most difficult problem with alkanes is to stop the oxidation at the stage of a necessary product, since oxidation of gaseous hydrocarbon on a solid surface tends to proceed up to the stable final products CO and CO$_2$ (a so-called deep oxidation). In deep oxidation the reactivity of alkanes, particularly those having a small number of carbon atoms, is usually lower than that of other compounds, e.g., olefins and acetylenes.

Methane is the least reactive among the alkanes, reaction rate generally increasing with the number of carbon atoms. Thus, the reaction rate on Cu_2O changes in the following order [17]:

Alkane:	CH_4	C_2H_6	C_4H_{10}	C_5H_{12}	C_6H_{14}	C_7H_{16}	C_8H_{18}
Rate $\times\ 10^{15}$ mole $m^{-2}s^{-1}$	4–8	9	51	79	108	120	144

Among the catalysts based on oxides, chromites, ferrites, cobaltites, and those containing cobalt as a cation are particularly good catalysts for methane oxidation at $300°C$ as compared with Mn, Fe, Ni, Cu, Mg, Cd [18].

Kinetic isotope effect measurements show that the rate controlling step for alkane oxidation involves C—H bond cleavage. The detailed mechanism is difficult to elucidate, since both the hydrocarbon molecule and dioxygen form various chemisorbed species on the surface.

Important results were obtained recently in the investigation of selective ammoxidation of short chain alkanes in the presence of ammonia.

Propane produces acrylonitrile on a gallium—antimony catalyst [19], while ethane forms saturated acetonitrile on molybdates of scandium and chromium [20]. Ammoxidation of butane also produces acetonitrile on vanadium—molybdenum oxide catalyst supported by silica. The large kinetic isotope effect in the case of ethane shows that the C—H bond cleavage is again the rate-determining step of hydrocarbon activation, possibly with the formation of a free radical.

Ammonia increases the rate of the reaction, apparently playing an important role in the activation mechanism. The possible mechanism of the initial hydrocarbon RH reaction on the catalytic surface may include electron transfer from RH to the oxidant molecule (Ox) with simultaneous H^+ abstraction by the molecule of the base (B)

$$Ox + RH + B \longrightarrow R\cdot + Ox^- + BH^+$$

We shall see later that such a mechanism is apparently common in the oxidation of alkanes in solution in the presence of high-valent metal complexes.

The activity of catalysts towards hydrocarbons may be considerably increased when transition metal compounds are used in zeolites. E.g., zeolite containing Cu(II) [Cu(II)Y] is more active in the oxidation of cyclohexane (to benzene, CO_2 and H_2O) than the usual catalyst of CuO supported on silica [21].

III.4. Reactions of Alkanes with Free Metal Atoms and Ions

Modern methods of investigation allow us to study the reactions of various compounds — including alkanes — with free metal atoms [22a]. For example, the co-condensation method was employed to study the reactions of zirconium atoms with isobutane and neopentane which co-condense at $77°C$ [22b].

Judging by the products formed, Zr atoms react via insertion into both the C—H and C—C bonds. Thus, for neopentane, the following reactions take place

$$Zr + \text{neopentane} \longrightarrow \begin{cases} HZrCH_2C(CH_3)_3 \\ H_3CZrC(CH_3)_3 \end{cases}$$

The reaction via the C—H bond to produce H—FeFe—CH_3 or FeFe—CH_3 [23] was discovered using Mössbauer spectroscopy in the reaction of Fe atoms and Fe_2 dimers in a matrix with CH_4 at $77°$ K. When no reaction takes place between ground state metal atoms and alkanes, photoexcitation often produces sufficiently active species to react with cleavage of the C—H bond. Thus irradiation with $\lambda < 360$ nm induces the reaction of iron atoms with methane producing CH_3FeH (infrared spectra). Similar results were obtained with Mn, Co, Cu, Zn, Ag, Au (formation of CH_3MH in all the cases), whereas no reaction was detected with Ca, Ti, Cr, Ni [24]. Metal atoms rather than dimers have been suggested to be active species in the reaction. A prolonged irradiation produces $(CH_3)_2M$.

Alkanes were found to react easily with transition metal ions in the gas phase. The results of ion beam studies of the reaction of Co^+ with alkanes [25] suggest that oxidative addition of both the C—H and C—C bonds are important reaction steps

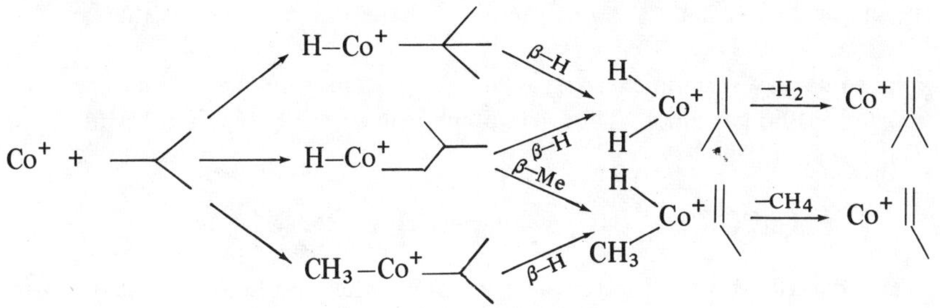

Fe^+ ions were shown to react with *iso*-butane according to the schemes [26]

$$Fe^+ + i\text{-}C_4H_{10} \longrightarrow \begin{cases} \xrightarrow{85\%} FeC_3H_6^+ + CH_4 \\ \xrightarrow{15\%} FeC_4H_8^+ + H_2 \end{cases}$$

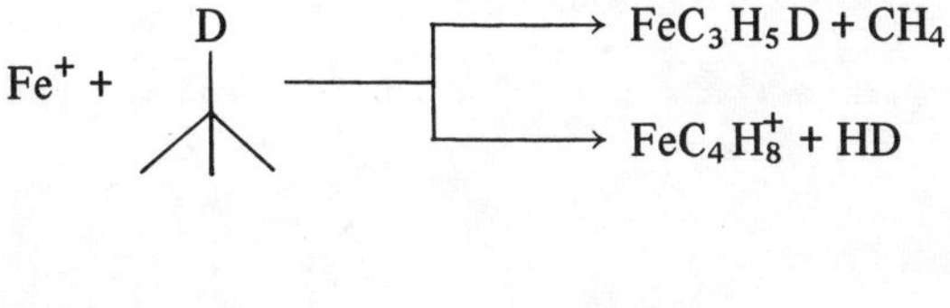

The following mechanism of hydrogen and methane formation may be suggested

$$Fe^+ \; + \; \rangle\!-\!\langle \;\longrightarrow\; CH_3-Fe^+ \;-\!\langle^{H} \;\longrightarrow$$

$$\longrightarrow\; \underset{CH_3}{\overset{H}{>}}Fe^+\cdots\| \;\longrightarrow\; Fe^+\cdots\| \;+\; CH_4$$

$$Fe^+ \; + \; \rangle\!-\!\langle \;\longrightarrow\; H-Fe^+ \;-\!\!\!\!\overset{H}{\underset{|}{\diagup}} \;\longrightarrow$$

$$\longrightarrow\; \underset{H}{\overset{H}{>}}Fe^+\cdots\| \;\longrightarrow\; Fe^+\cdots\| \;+\; H_2$$

Some difference between Fe^+ and Ti^+ was detected in [27] dealing with the gas-phase reaction of alkanes with free ions. While C—C insertion, leading to alkane elimination, and C—H insertion, leading to H_2 elimination, are competitive in the Fe^+–alkane system, the Ti^+–alkane system is dominated by C—H insertion.

In some cases complexes of alkanes with transition metal ions have been reported in the gas phase. Thus *iso*-butane reacts with $Fe(CO)^+$ substituting the CO molecule [28]

$$Fe(CO)^+ + i\text{-}C_4H_{10} \;\longrightarrow\; FeC_4H_{10}^+ + CO$$

The data in this fascinating field are still rather too scarce to make general conclusions, but one may assume that the main distinctive feature of these species is their apparent insensitivity to steric factors, allowing free atoms and ions to attack both the C—H and C—C bonds.

Hydrocarbon reactions at different metal centers (metal atoms, metal surfaces, mono- and polynuclear metal complexes) have been reviewed recently [29].

References

1. C. Kemball: *Advances in Catalysis*, p. 223, Acad. Press, Inc., New York and London (1959).

QB0I AIC3

2. R. L. Burwell: *Catal. Rev.*, 7, 25 (1972).
3. B. A. Kazanskii and A. L. Liberman: *Neftekhimija*, 12, 147 (1972).
4. A. Ozaki: *Isotopic Studies of Heterogeneous Catalysis*, Kadansha, Ltd. – Acad. Press, Tokyo, New York, San Francisco, London (1976).
5. D. A. Dowden: 'The Reactions of Hydrocarbons on Multimetallic Catalysts', *Catalysis*, 2, 1 (1978).
6. H. A. Quinn, J. H. Graham, and M. A. Kerney: *J. Catal.*, 22, 35 (1971).
7. J. Barbier, A. Morales, and R. Maurel: *Nouv. J. Chim.*, 4, 223 (1980).
8. V. Ponec and W. M. H. Sachtler: *J. Catal.*, 24, 250 (1972).
9. C. Kemball: *Proc. Roy. Soc.*, A217, 376 (1953).
10. S. Carrà and L. Forni: *Catal. Rev.*, 5, 159 (1971).
11. B. K. Skarchenko: *Usp. khim.*, 40, 2145 (1971).   QDEOI AIJ6
12. P. Biloen, F. M. Dautzenberg, and W. M. H. Sachtler: *J. Catal.*, 50, 77 (1977).
13. V. Ami-Ebrahimi, A. Choplin, P. Parayre, and F. G. Gault: *Nouv. J. Chim.*, 4, 431 (1980).
14. J. R. H. Van Schaik, R. P. Dessing, and V. Ponec: *J. Catal.*, 38, 273 (1975).
15. G. A. Somorjai: *Acc. Chem. Res.*, 9, 248 (1976).
    G. A. Somorjai: *Pure and Appl. Chem.*, 50, 963 (1978).
16. L. Ya. Margolis: *Okislenije uglevodorodov na geterogennykh katalizatorakh* (Oxidation of hydrocarbons by heterogeneous catalysts), Khimija, Moscow (1977).
17. A. G. Anshits, V. D. Sokolovskii, and G. K. Boreskov, A. A. Davydov, A. A. Budneva, V. I. Avdeev, and I. I. Zakharov: *Kinet. kataliz*, 16, 95 (1975).
18. V. V. Popovskii: *Thesis*, Tomsk University, Tomsk (1973).
19. Z. G. Osipova and V. D. Sokolovskii: *Kinet. kataliz*, 20, 510 (1979);
    idem: ibid, 20, 512 (1979);
    idem: ibid, 20, 910 (1979).
20. V. D. Sokolovskii and S. M. Aliev: *Kinet. katal.*, 21, 691 (1980);
    S. M. Aliev, V. D. Sokolovskii, and N. S. Kotsarenko: *ibid*, 21, 686 (1980);
    S. M. Aliev and V. D. Sokolovskii: *ibid*, 21, 971 (1980).
21. D. B. Tagiev and Kh. M. Minachev: *Usp. khim.*, 50, 1929 (1981).
22. a. G. A. Ozin and A. Vander Voet: *Acc. Chem. Res.*, 6, 313 (1973).
    b. R. J. Remick, T. A. Asunta, and P. S. Skell: *J. Amer. Chem. Soc.*, 101, 1320 (1979).
23. P. H. Barrett, M. Pasternak, and R. G. Pearson: *J. Amer. Chem. Soc.*, 101, 222 (1979).
24. W. E. Billups, M. M. Konarski, R. H. Hauge, and J. L. Margrave: *J. Amer. Chem. Soc.*, 102, 7393 (1980).
25. P. B. Armentrout and J. L. Beauchamp: *J. Amer. Chem. Soc.*, 102, 1736 (1980).
26. J. Allison, R. B. Freas, and D. P. Ridge: *J. Amer. Chem. Soc.*, 101, 1332 (1979).
27. G. D. Byrd, R. C. Burnier, and B. S. Freiser: *J. Amer. Chem. Soc.*, 104, 3565 (1982).
28. R. B. Freas and D. P. Ridge: *J. Amer. Chem. Soc.*, 102, 7129 (1980).
29. E. L. Muetterties: *Chem. Soc. Revs.*, 11, 289 (1983).

CHAPTER IV

HOMOGENEOUS OXIDATION OF ALKANES IN THE
PRESENCE OF METAL COMPOUNDS

Metal compounds may participate in the oxidation of hydrocarbons, including alkanes, in reactions of three main types:

1. Direct oxidation under the action of a metal compound in high oxidation state (without the participation of dioxygen).

2. Oxidation by dioxygen, peroxide and other oxidants with a metal compound playing the role of an initiator or a catalyst.

3. Coupled oxidation where a metal compound in low oxidation state induces stoichiometric hydrocarbons oxidation when being oxidized.

The mechanisms of these three reaction types often have much in common; moreover, one type of oxidation may change to another. For example, in the presence of an additional electron donor, stoichiometric oxidation of hydrocarbons of type 3 might become a catalytic oxidation coupled with the oxidation of the donor if the transition metal compound is involved in the oxidation-reduction cycle.

As for the elementary mechanism of hydrocarbon reactions in the oxidation of all three types, the metal compound may interact with the hydrocarbon molecule directly, or by forming active species (e.g., free radicals) which would react with the hydrocarbon independently of the metal compound itself.

IV.1. Direct Oxidation of Alkanes by Metal Compounds and Complexes in High Oxidation State

It has been mentioned above that a number of high-valent metal complexes can react with aromatic and alkylaromatic hydrocarbons. Alkanes are also able to enter into such reactions in comparatively mild conditions, though usually at a lower rate. A common factor in these reactions is the catalytic action of protic and aprotic acids, which is observed both for aromatic and aliphatic hydrocarbons.

Oxidants capable of interacting with alkanes may contain metal ions with various electronic configurations: d^0 (Cr^{VI}, Mn^{VII}, Ce^{IV}, Pb^{IV}); d^6 (Co^{III}); d^7 (Pt^{III}); or d^8 (Pd^{II}, Cu^{III}, Ag^{III}). The standard redox potential for ions reacting with alkanes with appreciable rate in aqueous solutions should not be less than 1.5 V. However, ions with even smaller redox potentials (as measured in aqueous solutions) may become active towards hydrocarbons in strongly acidic media (see below).

In most cases free radicals are evidently formed as intermediate species in reactions with alkanes. For branching hydrocarbons the 'normal' activity of C—H hydrocarbon bonds ($3° > 2° > 1°$) is usually observed. Hence it is particularly interesting that there exist some exceptions to that rule, where tertiary C—H bonds in alkanes display, by contrast, a very low activity. Such peculiarity is observed for some reactions with Co(III) compounds, for biological oxidation of alkanes with the participation of iron complexes, and also for reactions of alkanes with Pt(II) complexes where metal-carbon bonds are formed as intermediates.

IV.1.1. ALKANE OXIDATION BY COMPOUNDS OF CHROMIUM(VI) AND MANGANESE(VII)

Cr(VI) compounds — chromic acid, chromyl chloride and others — are traditional oxidants in organic and inorganic chemistry [1, 2]. The conversion of toluene to benzaldehyde by chromyl chloride, known as the Etard reaction [3], was discovered 100 years ago. The oxidative properties of chromic acid are revealed only in strongly acidic solutions and almost completely disappear in alkaline ones. In the process of oxidation, chromic acid, as a rule, is reduced finally to Cr(III), since Cr(V) and Cr(IV) are unstable in normal conditions.

Hydrocarbons are generally oxidized at higher temperatures than such compounds as alcohols and aldehydes. The alkanes show, as usual, the least activity, the primary C—H bond being the most stable in the oxidation. Hydrocarbons with tertiary C—H bonds form tertiary alcohols, whereas ketones are formed in the case of secondary methylene groups.

For normal paraffins the reaction rate is proportional to the number of CH_2 groups in the molecule [4]. The rate is usually also proportional to the acidity, indicating that the participation of protonated particles in oxidation (supposedly $O=Cr(OH)_3^+$ or $HCrO_3^+$). Initially, the reaction seems to proceed via H atom elimination by the oxygen atom attached to chromium:

$$RH + O=Cr(OH)_3^+ \longrightarrow R\cdot + HO—Cr^V(OH)_3^+$$

However, most of the radicals formed evidently do not get into the solution in a free state. This conclusion is supported, in particular, by a considerable retention of configuration in the oxidation of optically active hydrocarbons. For example, the chromic acid oxidation of (+)—3-methylheptane in 91% acetic acid gives (+)-3-methyl-3-heptanol with 70–85% retention of configuration [5]. The kinetic isotope effect (k_H/k_D) for 3-ethylpentane is 3.1, whereas for diphenylmethane $k_H/k_D = 6.4$.

The ratio of rates for chromic acid oxidation of primary, secondary and tertiary C—H bonds is similar to those of bromination. Therefore, the authors of [5] assume hydrogen atom abstraction to be a primary reaction of the oxidant with the alkane. The radical formed in the solvent 'cage' can give chromium(IV) ester (which then leads to alcohol formation with retention of configuration) or diffuse out of the cage racemic products.

$$R_3CH + Cr^{VI} \longrightarrow [R_3C \cdot Cr^V] \longrightarrow R_3C \cdot \xrightarrow{Cr^{VI}} R_3C^+ \xrightarrow{H_2O} R_3COH$$
$$\text{racemization}$$

$$\downarrow \text{recombination}$$

$$R_3COH \xleftarrow{\text{solvolysis}} R_3C-O-Cr^{IV} \xrightarrow{\text{hydrolysis}} R_3COH$$
$$\text{partial retention} \qquad\qquad\qquad\qquad \text{retention}$$

Along with the suggestion of the free radical interaction of chromic acid with the alkanes there have been some other hypotheses. For example, a resonance hybrid of the complexes corresponding to H atom and hydride abstraction was suggested as an intermediate in the two-electron oxidation [6]:

$$\text{>C<} \quad \overset{H}{\underset{O}{}} \quad + \quad \overset{HO}{\underset{}{=}} Cr^{VI} \overset{OX}{\underset{OH}{}} \longrightarrow$$

$$-\overset{H}{\underset{|}{C}} \cdot \quad O-Cr^V O_3 H_2 X \longrightarrow -\overset{H}{\underset{|}{C^+}} \quad O-Cr^{IV} O_3 H_2 X$$

$$[-\overset{|}{\underset{|}{C}}-O-Cr^{IV} O_3 H_3 X]^+ \longrightarrow -\overset{|}{\underset{|}{C}}-OH$$

X = H or COCH₃

Potassium permanganate is one of the best known oxidants of organic and inorganic compounds [1]. Here, as in the case with chromic acid, the oxidative properties are strengthened in acidic media but, as distinct from Cr(VI), permanganate is also a strong oxidant in both neutral and alkaline media.

Permanganate in solution of trifluoroacetic acid interacts with alkanes at room temperature [7]. The reaction rate changes linearly with the increase of acidity, h_R (indicating that active species are formed by protonation and subsequent water elimination), which implies that the MnO_3^+ cation, free or bound in a complex, is an active species. For isopentane the reaction rates of primary, secondary and tertiary C—H bonds correspond to the ratio $1 : 60 : 2100$. The rate calculated per methylene group in normal alkanes increases with an increase of chain length. Among cycloalkanes cyclohexane is the least reactive and the isotope effect (the rate ratio of C_6H_{12} and C_6D_{12} oxidation) is 4.3 at $25°C$.

The active species is a strong electrophile. This is proved by the inhibiting effect of electronegative substituents present in a hydrocarbon molecule: in 80% wt. trifluoroacetic acid, where propane and other alkanes are readily oxidized, practically no reaction is observed with nitroethane or propanone nitrile. Besides, alkylaromatic hydrocarbons are attacked by MnO_3^+ in the benzene ring, the effect of substituents being described by the equation

$$\log\frac{k_s}{k_0} = \rho^+ \sigma$$

with $\rho^+ = -5.29$. Accordingly, no isotope effect is observed if hydrogen is replaced by deuterium in the side chain (the reaction rates of $C_6H_5CH_3$, $C_6H_5CD_3$, and $C_6D_5CD_3$ are practically equal). Hence, MnO_3^+ is a stronger electrophile than such oxidants as Mn(III), Ce(IV) or Cr(VI) which attack alkylaromatic hydrocarbon derivatives preferentially on the side chain. The electrophilic properties are due, to a large extent, to the acid action, since permanganate in aqueous solutions also reacts with the α-C—H bond of the side chain of the aromatic ring.

IV.1.2. COMPLEXES OF RUTHENIUM(IV) AND IRIDIUM(IV)

Ruthenium(IV) chloride complexes considerably enhance alkane oxidation by aqueous solutions of chromic acid [8, 9]. Primarily chloroalkanes are formed. The dependence the reaction rate of propane and butane on the concentration of chloride ions reaches its maximum at about 0.5M Cl^- concentration. The selectivity of the C—H bond attack determined by the ratio of reaction rates is [8]:

$$1° : 2° : 3° = 1 : 10^2 : 10^4$$

which is very close to the selectivity of non-catalytic alkane oxidation by chromic acid in sulfuric acid.

At $[Cl^-] > 0.5M$ the reaction rate depends only weakly on the acidity and ionic strength, whereas at $[Cl^-] < 0.5M$ the rate is greatly increased with an increase of the acidity. The following reaction mechanism has been proposed [9]:

$$RH + Ru(IV) \rightleftharpoons Ru(IV)RH$$
$$Ru(IV) \cdot RH + Cr(VI) \longrightarrow products$$

the nature of the hydrocarbon complex with ruthenium(IV) being still un-
specified.

Ruthenium(IV) complexes also catalyze the oxidation of alkanes by manga-
nese(III) in sulfuric acid solutions [10]. The reaction is first order with respect
to ruthenium concentration in the region of $10^{-5}-10^{-6}$ M. The dependence
of the rate constant on the concentration of manganese(III) passes through
a maximum. The selectivity of the oxidation of alkanes, as well as other characteristics of reactions in the systems Mn(III)$-$H$_2$SO$_4$ and Mn(III)$-$Ru(IV)$-$
H$_2$SO$_4$, is very similar (see Section IV.1.4).

Iridium(IV) complexes also catalyze the oxidation of alkanes by chromic acid
[11, 12]. The reaction is first order with respect to hydrocarbon concentration.
The following ratio is observed for the selectivity of the C$-$H bond attack (for
the cases of ethane, propane, n-butane and iso-butane):

$$1° : 2° : 3° = 1 : 30 : 250$$

i.e. the selectivity here differs greatly from that observed in the chloride catalyzed
Ru(IV) reaction, showing a relatively higher activity of the primary C$-$H
bond.

The ratio between rate constants of isobutane oxidation in the systems:
Cr(VI)$-$RH **(1)**, Ir(III)$-$Ir(H$_2$O)Cl$_5^-$$-$RH **(2)**, IrCl$_6^{2-}$$-$Cr(VI)$-$RH **(3)**, and
Ir(H$_2$O)Cl$_5^-$$-$Cr(VI) $-$RH **(4)** is $k_1 : k_2 : k_3 : k_4 = 0.3 : 0.7 : 10 : 100$ ($t =$
97.5°C), which demonstrates that Ir(H$_2$O)Cl$_5^-$ is about one order of magnitude
more catalytically active than IrCl$_6^{2-}$.

The authors of reference [11] attribute this to the stronger (in comparison
with IrCl$_6^{2-}$) electrophilic properties of Ir(H$_2$O)Cl$_5^-$ and the increase of redox
potential in substituting chloride ions for water molecules in the coordination
sphere of iridium.

The rate of alkane oxidation in systems **3** and **4** increases linearly with
acidity ([H$_2$SO$_4$] = 0.1$-$1.33 M) and drops sharply with rising Cl$^-$ concentration. The authors of [11, 12] explain this acidic action in terms of acid
suppressing the dissociation

$$\text{Ir(H}_2\text{O)Cl}_5^- \;\rightleftharpoons\; [\text{Ir(OH)Cl}_5]^{2-} + \text{H}^+$$

and the Cl$^-$ action probably by the formation of a less active oxidant form:

$$\text{H}^+ + \text{Cl}^- + \text{HCrO}_4^- \longrightarrow \text{ClCrO}_3^- + \text{H}_2\text{O}$$

The catalytic role of ruthenium(IV) and iridium(IV) compounds might
be connected with a particular reactivity of high-valent Ru(IV) and Ir(IV)

compounds towards the alkanes. However, the data obtained up to the present time are not sufficient to make a decisive conclusion on the mechanism of these reactions. The absence of marked changes in the system's selectivity upon introduction of ruthenium(IV) complexes raises some doubt, at least as for ruthenium(IV) playing a direct role in the redox process. There is a strong possibility that highly charged ruthenium complexes participate in an electrophilic coaction in the reactions of chromic acid [12] (as do Al^{3+} ions in the reaction of Pd(II) with alkanes in $H_2 SO_4$) (see Section IV.1.4).

IV.1.3. PECULIARITIES OF ALKANE OXIDATION BY COMPOUNDS OF COBALT(III). OTHER OXIDANTS

The oxidation of organic compounds by cobalt(III) salts has been studied in detail in many papers (see reviews [13–15]); the oxidation of aromatic and alkylaromatic compounds has already been mentioned above. The oxidation of saturated hydrocarbons including normal, iso- and cyclic alkanes also takes place, both in the presence and absence of dioxygen.

Cobalt(III) perchlorate oxidizes cyclohexane as well as aromatic and alkyl-aromatic hydrocarbons in aqueous acetonitrile at room temperature. The reaction is first order with respect to both the substrate and oxidant [16].

The reactivity of the compounds studied corresponds to the following order:

$$p\text{-NO}_2 C_6 H_4 Me < C_6 H_{12} < PhCH_2 CH_2 Ph \sim PhMe \sim PhEt < Ph_2 CH_2 <$$

$$< p\text{-Bu-}C_6 H_4 Me < Ph_2 < \text{naphthalene} < \text{anthracene, phenanthrene}.$$

The authors [16] have come to the conclusion that purely aromatic hydrocarbons react with cobalt(III) ions by donating one electron to the oxidant from their π-orbital systems; while alkylaromatic hydrocarbons, such as toluene and ethylbenzene, interact preferentially by hydrogen transfer of the aliphatic α-C—H group. If this H atom transfer is not the initial stage of the reaction, then proton abstraction must proceed synchronously with electron transfer. This is proved by a readily occuring reaction with cyclohexane, in which an electron transfer to form an ion-radical is practically improbable (see Section IV.3). When the oxidation is effected by cobalt(III) trifluoroacetate in trifluoroacetic acid at 30°C, cyclohexane is converted into cyclohexyl trifluoroacetate within 23 hours with 35% yield [17]. It is interesting to note that in this case the reaction of benzene to produce phenyltrifluoroacetate occurs within 0.5 h at room temperature, toluene reacting under the same conditions even at −16°C. Thus the reaction rate (according to [17]) in trifluoroacetic acid for different hydrocarbons increases with decreasing ionization potential (being 9.88, 9.25, and 8.82 eV for cyclohexane, benzene, and toluene, respectively).

The reaction kinetics of hydrocarbon oxidation by Co(III) complexes is usually described by the equation

$$-\frac{d[Co(III)]}{dt} = \frac{k[RH]\,[Co(III)]^2}{[Co(II)]}$$

In trichloroacetic acid the main products of heptane oxidation in the absence of dioxygen are heptyl chlorides and heptyl acetates [18], whereas under an oxygen atmosphere heptanones are mostly formed.

The important feature of these reactions is their unusual selectivity in reactions with primary, secondary and tertiary C—H bonds. Almost exclusively, secondary C—H groups are attacked in the case of linear alkanes, yielding the products mainly at the second atom of the hydrocarbon chain. Thus, oxidation of heptane in an inert atmosphere in trifluoroacetic acid produces an 85% yield of acetates, 81% of this being in the form of 2-acetoxyheptane. Under oxygen in trichloroacetic acid, ketones reach 80% yield, 2-heptanone making up about 83% of the ketones formed [18].

The authors of [18] suggest the following reaction mechanism:

$$RH + Co(III) \underset{}{\overset{k_1}{\rightleftharpoons}} R\cdot + Co(III)$$

$$R\cdot + Co(II) \xrightarrow{k_2} products$$

An intermediate complex involving an alkyl cobalt derivative species is assumed to be formed here

$$RH + Co(III) \rightleftharpoons RCo(IV) \rightleftharpoons R\cdot + Co(II)$$

Formation of heptyl acetates in heptane oxidation may be explained by the following scheme:

$$R\cdot + Co(III) \longrightarrow R^+$$

$$R^+ + AcOH \longrightarrow AcOR + H^+$$

The formation of heptyl acetate is also possible when an acetate ion as a ligand at cobalt(III) is attacked by radicals with a simultaneous electron transfer to the Co^{3+} ion. Alkyl chlorides are evidently formed upon the elimination of a chlorine atom of trichloroacetic acid by the radical:

$$R\cdot + CCl_3CO_2H \longrightarrow RCl + \cdot CCl_2CO_2H$$

An unusual selectivity in the alkane reaction with Co(III) is revealed, in particular, by the fact that, in the case of alkanes with tertiary C—H bonds, their reactivity is low. For the oxidation of 2-methylpentane by cobalt(III) acetate in the presence of trichloroacetic acid under dinitrogen the distribution of isomeric chlorides turned out to be as follows:

1–6%; 2–5%; 3–2%; 4–74%; 5–13%.

A significant prevalence of the isomer formed in the attack of the secondary C—H bond of the CH_2 group of the fourth carbon atom over the one formed from the tertiary C—H bond, whose dissociation energy is far lower, may be evidently understood as an effect of steric hindrance.

The study of some Co(III) acetate complexes has led to the conclusion that they have a complex polynuclear structure [19, 20], which is in agreement with the possibility of strong steric hindrances in the reaction with hydrocarbons.

In the case of the reaction of adamantane with cobalt(III) acetate in acetic acid in the presence of trifluoroacetic acid a hydrocarbon acetoxylation occurs, whereas chlorination takes place in the presence of carbon tetrachloride. The tertiary C—H bond, which is not very sterically hindered in adamantane, is attacked in this case [21]. High yields of 1-adamantyl trifluoroacetate are also observed in adamantane oxidation by manganese(III) and lead(IV) acetates in trifluoroacetic acid [22].

The comparison of anodic oxidation where a cleavage of the C—C bond and molecular fragmentation are observed with selective adamantane oxidation effected by Pd(IV), Co(III), Mn(III) leads the authors to believe that cation-radicals are not formed in chemical oxidation. The proposed mechanism includes localized attack by the electrophilic reagent on the tertiary C—H bond.

As shown in [23], manganese(III) acetate acts more selectively than cobalt(III) acetate (a higher yield of cyclohexyl acetate is obtained), which is explained by its low reactivity. Lead(IV) trifluoroacetate oxidizes aromatic and alkylaromatic hydrocarbons into esters of trifluoroacetic acid, even at room temperature [24]. Heptane under these conditions is less reactive than benzene. Adamantane and its analogues are oxidized by lead(IV) trifluoroacetate also at 20°C in a mixture of methylene chloride and trifluoroacetic acid to give high yields of tertiary alcohols [25]. In this case the reaction is strongly catalyzed by Cl^- ions. As was indicated above, the Pb(IV) reaction with arenes occurs via the mechanism of electrophilic substitution, which is much less likely for alkanes.

Copper(III) complexes with periodate oxidize cyclohexane in aqueous solution at room temperature [26] to give cyclohexanol, cyclohexanone and benzene. The yield of products is low due to the fast oxidation of cyclohexanol and cyclohexanone by copper(III) complexes with periodate at room temperature.

There is an evidence [27] of silver(III) salts reacting with cyclohexane.

IV.1.4. OXIDATION OF ALKANES IN CONCENTRATED SOLUTIONS OF SULFURIC ACID

Since 1973 Rudakov and coworkers in a series of papers (see reviews [10, 28])
have reported that saturated hydrocarbons are oxidized by some metal complexes in concentrated solutions of sulfuric acid. H_2SO_4 increases the oxidative
ability of the complexes so as to permit the reaction with alkanes to proceed,
even at 25–100°C. Solutions of palladium(II), platinum(III), manganese(III)
or mercury(II) may be used as oxidants.

The data on reaction products are still very scarce; olefins and carbenium
ions, R^+, are believed to be produced as intermediates.

The participation of metal complexes in the reaction of alkanes is not crucial.
Other oxidants containing no metals may also be active towards the alkanes
in sulfuric acid. They are: hydrogen peroxide, ammonium persulfate, and nitric
acid. Sulfuric acid itself can also oxidize alkanes when highly concentrated.
Evidently, in these cases, the active species are such electrophiles as NO_2^+ (in the
case of HNO_3) or HSO_3^+ (in the case of sulfuric acid itself).

Metal ions, if they are not sufficiently strong oxidants, e.g., copper(II),
iron(III), aluminium(III), nickel(II), silver(I), rhenium(III), rhodium(III) compounds, molybdates, tungstates, are inactive with respect to the alkanes in
sulfuric acid. Therefore, metal complexes in sulfuric acid behave like oxidants
strengthened by acid electrophilic properties. Standard one- or two-electron
redox potentials of oxidants active in sulfuric acid have to be not lower than
~0.9 V (when measured in aqueous solution), whereas as has been mentioned
above, the oxidants active towards alkanes in water must have potentials not
lower than ~1.5 V.

Table IV.1 gives some characteristics of oxidants in sulfuric acid and, for
comparison, of some other systems, reacting with alkanes.

The following features are common for oxidants (metal and non-metal)
acting in sulfuric acid [10]:

1. The reaction follows a second-order kinetic equation (except manganese(III)
complexes where species active to alkane are obviously the radicals produced
in manganese complex decomposition which occurs independently of the
alkanes)

$$- \frac{d[RH]}{dt} = k_2 [RH] [M^{n+}]$$

2. In the reaction rate-controlling step the C—H bond in the alkane is cleaved.
The kinetic isotope effect, k_H/k_D, for almost all the systems studied is 2 ± 0.2
and is the same for the cleavage of the *tert*- and *sec*-C—H bonds.

TABLE IV.1

Systems of homogeneous oxidation of alkanes and some characteristics of selectivity

Oxidant system	k_H/k_D**	$\dfrac{c\text{-}C_6H_{12}}{c\text{-}C_5H_{10}}$	$3° : 2°$	$-\rho*$	δ
Hg(II)–H$_2$SO$_4$	2.0	3	3000	6.8	1.3
Pt(III)–H$_2$SO$_4$	1.9	2.3	1300	6.2	1.5
H$_2$SO$_4$	2.0	2.2	960	6.0	0.8
Pd(II)–H$_3$PO$_4$–BF$_3$	–	2	700	–	–
NO$_2^+$–H$_2$SO$_4$	2.1	3.1	85	3.9	0.5
Pd(II)–H$_2$SO$_4$	2.0	2.8	94	3.6	0.9
Cr(VI)–H$_2$SO$_4$	5.2	0.7	60	3.3	0.4
HNO$_3$–Ru(IV)–Cl$^-$–H$_2$O	2.6	1.3	35	3.2	0.6
Cr(VI)–Ru(IV)–Cl$^-$–H$_2$O	4.3	0.8	24	3.3	0.7
H$_2$O$_2$–H$_2$SO$_4$	2.1	1.7	24	2.8	0.5
Mn(III)–Ru(IV)–H$_2$SO$_4$	–	1.2	20	2.7	0.9
Mn(III)–H$_2$SO$_4$	1.9	1.6	12	1.7	0.3

** $c\text{-}C_6H_{12}/c\text{-}C_6D_{12}$ or $(CH_3)_3CH/(CH_3)_3CD$

$\rho*$ and δ are parameters of the Taft equation (see below).

For some of the oxidants, such as CrO_3, MnO_3^+, ozone (O_3–CCl_4), k_H/k_D is higher and reaches 4 ± 1. It is of interest that, for these oxidants, the ratio of reaction rate constant of cyclohexane to that of cyclopentane is less than unity, i.e. the reaction rate follows the C—H bond energy change which decreases in the sequence

$$C_3 > C_4 > C_6 > C_5 > C_7 > C_8$$

while for other oxidants in Table IV.1 the reaction rate follows the ionization potential change, which is lower for C_6H_{12} than for C_5H_{10}.

Activation energy decreases with the decrease of the C—H bond energy (Table IV.2).

3. In all the systems the selectivity of the C—H bond cleavage follows the order: $3° > 2° > 1°$, which corresponds to the rate decrease with increasing C—H bond energy. This selectivity is different for different reagents; the ratios $3° : 2°$ and $2° : 1°$ decrease in the order:

$$Hg(II) > Pt(III) > H_3SO_4^+ > NO_2^+ > Pd(II) > S_2O_8^{2-} \simeq Mn(III)$$

For the most selective system, Hg(II)–H$_2$SO$_4$, $3° : 2° = 3000$; for the least selective, Mn(III)–H$_2$SO$_4$, $3° : 2° = 12$.

TABLE IV.2

Activation parameters of the oxidation of alkanes in systems $M^n - H_2SO_4$

Hydrocarbon	Temperature range, °C	H_2SO_4 %	E_A^* kcal mole^{-1}	$\log A$
$H_3SO_4^+$-sulfuric acid				
Methylcyclohexane	80–98	94.9	20	9.7
Ethylcyclohexane	75–90	97.7	20	10.2
2-Methylpropane	70–99	97.7	21	10.5
Pd(II)-sulfuric acid				
Cyclohexane	43–90	94.9	17	10.6
Methylcyclohexane	45–90	94.9	15	9.7
NO_2^+-sulfuric acid				
Cyclohexane	15–50	93.0	16	11.9
n-Octane	15–65	93.0	14.6	10.0
2,2,4-Trimethylpentane	25–65	93.0	15	10.6
2-Methylpropane	25–60	93.0	11	8.0
2-Methylpropane	25–65	96.6	11.9	7.4
$S_2O_8^{2+}$-sulfuric acid				
Cyclohexane	7–37	93.0	13.9	8.1
Methylcyclohexane	10–40	93.0	12.9	7.8
Propane	35–65	95.4	13.9	7.8
2-Methylpropane	20–50	95.4	12.9	8.2

* Errors in E_A ca. 1 kcal mole^{-1}

4. For the alkanes with a *tert*-C—H bond the rate increases in the order

$$Me_3CH < Me_2EtC—H < MeEt_2C—H < Et_3C—H$$

in accordance with an increase of electron-donor ability of alkyl groups. For all the systems of the $M^n - H_2SO_4$ type the effect of substituents follows the Taft equation

$$\log k = \text{const} + \rho^* \Sigma \sigma^* + \delta \Sigma E_S^0$$

with the ρ^* value varying from -3 to -1, which is characteristic of radical II elimination. Steric hindrances exert a rather weak influence here.

5. The reaction rate increases exponentially with an increase of acid concentration following the Hammett equation for acid catalysts

$$\log k = \text{const} - mH_0$$

This indicates that protonation is a way in which activating species react with the alkanes.

6. Reviewing the results obtained the authors of [10, 28] noticed that the participation of bases is the common feature for all the systems reacting in sulfuric acid. For example, in the case of the NO_2^+–H_2SO_4 system the reaction rate reaches its maximum in 92–94% sulfuric acid. Rudakov and coworkers come to the conclusion that similarity in the systems' behaviour in sulfuric acid with respect to alkanes indicates a common mechanism for the limiting stage. They propose a homolytic oxidative C–H bond cleavage with the participation of a ligand L as the most likely mechanism

$$RH + L^-M^n \longrightarrow R\cdot + HL \ldots M^{n-1}$$

IV.1.5. PALLADIUM(II) COMPLEXES

Among systems studied by Rudakov and coworkers, the alkane reactions in a solution of palladium(II) sulfate in sulfuric acid were investigated most thoroughly. When palladium(II) concentrations are approaching 0.01 M, the reaction is first order with respect to palladium concentration. The Hammett equation

$$\log k = \text{const} - mH_0$$

holds for different H_2SO_4 concentrations with $m = 1$ for palladium.

Evidently, the protonated palladium(II) complex is an oxidizing agent in the reaction with alkanes. It follows from the kinetic data, that associates involving two or more palladium atoms acquire here an important role with palladium concentrations greater than 0.01M. In these associates palladium ions obviously exert electrophilic coaction in the palladium(II) interaction with the alkanes. In accordance with this suggestion the additions of other positive ions, e.g., Al^{3+}, act in the same way. Not being oxidants themselves they considerably increase the rate of oxidation of alkanes by palladium(II) in sulfuric acid.

Palladium(II) trifluoroacetate can oxidize an alkane (n-hexane or cyclohexane) in trifluoroacetic acid at 92°C [29]. Palladium itself is reduced to palladium(0), and cyclohexane is oxidatively dehydrogenated to benzene according to the stoichiometry:

$$3\,Pd(CF_3CO_2)_2 + C_6H_{12} \longrightarrow 3\,Pd(0) + C_6H_6 + 6\,CF_3CO_2H$$

In H_3PO_4–BF_3 medium (1:1), palladium(II) phosphate also reacts with the alkanes, the reaction kinetics having the same regularities as in the solution of sulfuric acid [10].

IV.2. Participation of Transition Metal Ions and Complexes in the Oxidation of Hydrocarbons by Molecular Oxygen

The oxidation of hydrocarbons by molecular oxygen includes very important chemical processes known for a very long time. Their importance is connected with the necessity of rational use of hydrocarbons from coal, oil and natural gas. Despite the great amount of work devoted to this problem, it is far from being completely solved. At present its significance, in terms of the urgent necessity of more economical consumption of natural resources, is increasing with time.

Thermodynamically the formation of oxygen-containing products from hydrocarbons and dioxygen is practically always favourable, since oxidation reactions are highly exothermic. However, it is this very fact that hinders the creation of selective processes, the difficulty usually lying in the prevention of different parallel and secondary oxidation reactions which lead to a variety of by-products.

The use of salts and complexes of transition metals creates great possibilities for solving problems of selective oxidation, as has been demonstrated for a number of important processes.

IV.2.1. CHAIN MECHANISM OF OXIDATION

To explain the molecular oxygen interaction with organic compounds Bach and Engler proposed the so-called peroxide theory at the end of the 19th century [30, 31]. In accordance with this theory the initial reaction products are peroxides, which are later transformed into more stable products.

Further investigations have confirmed the intermediate formation of peroxides. However, a real mechanism of autoxidation involving hydrocarbons was elucidated only with the discovery of chain reactions and the formation, mainly by Semenov and his coworkers, of the theory of branched chain processes [32].

A detailed study of liquid phase oxidation including that of alkanes for the last several decades has led to considerable understanding of the reaction mechanism (e.g., see [33]).

Catalytic and non-catalytic autoxidation of hydrocarbons (including the alkanes) is usually a branched chain process with a so-called 'degenerate' chain branching. This means that branching of each chain happens much later than its termination (as distinct from non-degenerate branching which occurs virtually simultaneously with chain propagation) and is caused by the formation of a rather stable intermediate which is nevertheless chemically more active than the initial hydrocarbon and can form free radicals at a greater rate than that of the chain initiation process. Hydroperoxides turned out to be the intermediates in liquid phase oxidation, thus confirming the peroxide theory. The comparatively low energy of the O—O bond in hydroperoxides brings about its cleavage with formation of free radicals.

The study of ROOH hydroperoxides produced from RH hydrocarbons has shown the structure of R in ROOH to be the same as in the initial RH, which confirms that a hydroperoxide is formed in the interaction of $RO_2 \cdot$ radical with the molecule of the initial hydrocarbon

$$RO_2 \cdot + RH \longrightarrow RO_2H + R \cdot$$

The evidence for the chain mechanism of hydrocarbon oxidation is primarily based on the observation of the enhancing effect of light, ionizing radiation and small additions of various initiators easily decomposing into free radicals, as well as the inhibiting effect of such compounds as phenols or aromatic amines that readily react with free radicals. For some oxidation reactions $RO_2 \cdot$ radicals involved in the chain propagation were directly observed by EPR.

The following classical scheme presents a typical mechanism of liquid-phase hydrocarbon oxidation for the early stages when the effect of reaction products may be neglected [33].

(0)* $RH + O_2 \xrightarrow{k_0} R \cdot + HO_2 \cdot$ Chain initiation

(1) $R \cdot + O_2 \xrightarrow{k_1} RO_2 \cdot$

(2) $RO_2 \cdot + RH \xrightarrow{k_2} RO_2H + R \cdot$ $\Bigg\}$ Chain propagation

(3) $ROOH \xrightarrow{k_3} RO \cdot + \cdot OH$ or

(3′) $2\,RO_2H \longrightarrow RO \cdot + RO_2 \cdot + H_2O$ $\Bigg\}$ Chain branching

(4) $R \cdot + R \cdot \longrightarrow RR$

(5) $RO_2 \cdot + R \cdot \longrightarrow ROOR$ Chain termination

(6) $RO_2 \cdot + RO_2 \cdot \xrightarrow{k_6} ROH + R'COR'' + O_2$

Each specific hydrocarbon has its own characteristics. However, in principle, the branching chain mechanism has some general features. The rate constants of reactions 1–6 of $R \cdot$ and $RO_2 \cdot$ radicals are high and their concentrations quickly reach stationary values. At sufficiently high dioxygen concentration $[RO_2 \cdot] \gg [R \cdot]$ and the termination proceeds only by reaction 6. For the steady-state conditions

$$\frac{d[R \cdot]}{dt} = v_i - k_1 [O_2] [R \cdot] + k_2 [RH] [RO_2 \cdot] = 0$$

$$\frac{d[RO_2 \cdot]}{dt} = k_1 [O_2] [R \cdot] - k_2 [RH] [RO_2 \cdot] - k_6 [RO_2 \cdot]^2 = 0$$

where v_i is the overall rate of radicals formation in chain initiation.

* The numbering of this radical-chain reaction scheme (and that on p. 84) is conventionally accepted in the literature.

Summing up both the equations we get

$$v_i = k_6 [RO_2 \cdot]^2 \qquad [RO_2 \cdot] = \left(\frac{v_i}{k_6}\right)^{\frac{1}{2}}$$

and the hydrocarbon oxidation rate at early stages where the rate of branching may be neglected will be

$$v = -\frac{d[RH]}{dt} = k_2 [RH] [RO_2 \cdot] = k_2 [RH] \left(\frac{v_i}{k_6}\right)^{\frac{1}{2}}$$

The low O—O bond energy causes the rate of radical formation in reactions 3 and 3' in the process of hydroperoxide accumulation to exceed the chain initiation rate, which is the reason for the total reaction rate increase during the process of hydroperoxide accumulation.

The reaction rate will stop increasing when the hydroperoxide accumulation rate via reaction 2 becomes equal to the rate of its decomposition into radicals via reactions 3 and 3'. The latter, in its turn, must be equal to the rate of radical termination via reaction 6.

Thus, the reaction rate will reach its maximum when

$$nk_2 [RO_2 \cdot] [RH] = 2k_6 [RO_2 \cdot]^2$$

and according to Walling [34] the maximum reaction rate will be the following

$$v_{max} = -\frac{d[RH]}{dt} = f^{-1} k_2 [RO_2 \cdot] [RH] = \frac{nk_2^2}{2fk_6} [RH]^2.$$

Here n is the average number of radicals produced per ROOH decomposed and f is the fraction of RH consumed which disappears by one ROO · attack. If the branching proceeds via reaction 3 ($n = 2, f = 1/3*$), then

$$v_{max} = -\frac{d[RH]}{dt} = \frac{3k_2^2}{k_6} [RH]^2.$$

If radicals are formed in bimolecular reaction 3' ($n = 1, f = 2/3$), then

$$v_{max} = \frac{3}{4} \frac{k_2^2}{k_6} [RH]^2.$$

This maximum rate in non-catalytic oxidation (if [RH] is the initial hydrocarbon concentration) is never really reached, due to a slow increase of hydroperoxide concentration and complications resulting from the accumulation of products.

* After the decomposition of ROOH molecule both RO · and · OH radicals react with RH in fast reactions, e.g. · OH + RH $\longrightarrow$ H$_2$O + R ·, hence per each reaction 2 three RH molecules disappear.

IV.2.2. CATALYTIC OXIDATION OF HYDROCARBONS IN THE PRESENCE OF SMALL CONCENTRATIONS OF TRANSITION METAL IONS

The catalytic effect of metal compounds in the oxidation of hydrocarbons and other organic compounds is a well-recognized fact. Salts of cobalt, manganese, iron, copper, chromium, lead, nickel are used as catalysts, which allow the reactions to be carried out at lower temperatures, therefore increasing the selectivity of oxidation. However, it is more important that the catalyst itself may regulate the selectivity of the process, leading to a formation of a particular product.

The studies of the mechanism of transition metal salt involvement have shown their role to consist, in most cases, in enhancing the formation of free radicals in the interaction with initial and intermediate species.

The common mechanism of catalysis by transition metal salts was formulated at the beginning of the thirties by Haber and Willstäter [35]. They propose a scheme involving metal ion interaction with a molecule (AB) involving a change of ion oxidation state and free radical formation:

$$M^{n+} + A—B \longrightarrow M^{(n+1)+} + A\cdot + B^-$$

$$M^{(n+1)+} + A—B \longrightarrow M^{n+} + A^+ + B\cdot$$

According to this scheme a metal ion alternatively increasing and decreasing its valency can participate in the formation of a large number of free radicals, thus playing the role of a catalyst.

The study of the participation of metal ions in the chain reactions of oxidation shows them to take part in all the stages of the reaction, i.e. chain initiation, branching, propagation, and termination.

In effect, small additions of transition metal complexes increase the initial rate of chain reactions and decrease the induction period, which is characteristic of non-catalytic oxidation. For example, naphthenates of cobalt, chromium and manganese reduce the induction period in cyclohexane oxidation. The same action is displayed by cobalt stearate, the induction period being shorter if more catalyst is introduced [33].

Direct measurements by a so-called inhibitor method show that the rate o. chain initiation in the absence of the catalyst is equal to $10^{-8} M^{-1} s^{-1}$, whereas the initiating rate reaches $4.7 \times 10^{-7} M^{-1} s^{-1}$ when cobalt stearate is introduced at a concentration of 0.06 mol.%.

Formation of free radicals in the presence of transition metal compounds may be the result of their interaction with hydrocarbon or dioxygen.

Such catalysts as cobalt and manganese salts are usually introduced in the bivalent state. An increase of chain initiation rate here may be connected with the activation of dioxygen by a metal complex via the following reaction [36]:

$$M^{2+} + O_2 \rightleftharpoons M^{(2+\delta)+} \cdots O_2^{\delta-}$$

$$M^{(2+\delta)+} \cdots O_2^{\delta-} + RH \longrightarrow M^{2+} + R \cdot + HO_2 \cdot$$

If the catalyst is in a high-valent state, chain initiation can proceed in the reaction, which has been already discussed above

$$M^{3+} + RH \longrightarrow M^{2+} + R \cdot + H^+$$

This mechanism was unequivocally established for the case of benzaldehyde oxidation at room temperature in solutions of glacial acetic acid [37, 38]. The kinetic equation obtained for the reaction is

$$v = k[RCHO]^{3/2} [catalyst]^{1/2}$$

As was seen (p. 76) for the chain termination of the second order, the chain reaction rate is

$$v = const. [RH] v_i^{1/2}$$

Hence,

$$v_i = k_1 [RCHO] [catalyst].$$

The rate constant of chain initiation was determined from the special experiments with the introduction of an inhibitor. It was found to be:

$$k = 3 \times 10^9 \exp(-14800/RT) \, M^{-1}s^{-1}.$$

The rate constant of the reaction of cobalt(III) acetate with benzaldehyde in the absence of dioxygen was determined in independent experiments. It turned out to be virtually the same as the rate constant of the chain initiation in the oxidation reaction in the presence of O_2.

However, the contribution of chain initiation to the radical formation is insignificant in the developed oxidation process, since the radicals are mainly formed in the reactions of intermediates in the process of degenerate chain branching.

These reactions are also catalyzed by transition metal ions. Especially well studied is the acceleration of radical decomposition of intermediately formed hydroperoxides.

As early as 1932 Haber and Weiss suggested a mechanism of interaction between hydrogen peroxide and iron(II) ions where electron transfer to the peroxide molecule led to the formation of a free hydroxyl radical [39]:

$$Fe^{2+} + H_2O_2 \longrightarrow Fe^{3+} + OH^- + OH\cdot$$

The reaction of iron(II) ions with hydroperoxides proceeds similarly [40]:

$$Fe^{2+} + ROOH \longrightarrow Fe^{3+} + OH^- + RO\cdot$$

Other ions, such as Co^{2+} and Mn^{2+} can react in the same way:

$$Co^{2+} + ROOH \longrightarrow Co^{3+} + OH^- + RO\cdot$$

The newly formed high-valent ions can further react with hydroperoxides, e.g.,

$$Co^{3+} + ROOH \longrightarrow RO_2\cdot + Co^{2+} + H^+$$

Therefore, transition metal salts, in particular those of cobalt, can catalyze the decomposition of hydroperoxides into free radicals:

$$2\,RO_2H \xrightarrow{\;Co^{II-III}\;} RO\cdot + RO_2\cdot + H^+ + OH^-$$

In this case the time taken to reach the maximum rate, as well as the steady-state concentration of hydroperoxide, will be reduced.

In non-polar hydrocarbon media the reaction of hydroperoxide proceeds with undissociated salt, for example, cobalt(II) stearate, which can be present in solution in small concentrations:

$$St_2Co + ROOH \longrightarrow RO\cdot + St_2CoOH$$

$$St_2CoOH + ROOH \longrightarrow RO_2\cdot + St_2Co + H_2O$$

Such a reaction obviously proceeds more slowly than in aqueous or other polar media, but much faster than does thermal hydroperoxide decomposition. This results in a sharp drop of steady-state concentration of the hydroperoxide in hydrocarbon oxidation in the presence of transition metal salts (Fig. IV.1). Here, a decrease of hydroperoxide concentration will be compensated by an increase of its specific rate of decomposition so that the maximum oxidation rate will remain the same as in non-catalytic oxidation. In effect, the maximum rate

$$v_{max} = \left(-\frac{d[RH]}{dt}\right)_{max} = f^{-1} k_2 [RO_2\cdot]_{max}[RH]$$

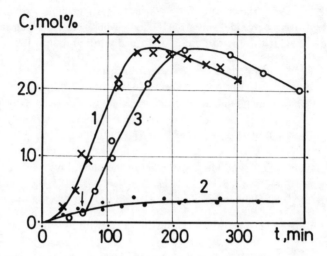

Fig. IV.1: The kinetics of accumulation of hydroperoxide in the oxidation of *n*-decane: (1) without a catalyst; (2) in the presence of 0.3 mole % St_2Co; (3) at the removal of the catalyst residue from the reaction area (the moment of removal is marked with an arrow); temperature is 140°C [33].

will be reached under the condition that

$$nk_2 [RO_2 \cdot] \cdot [RH] = 2k_6 [RO_2 \cdot]^2$$

that is

$$v_{max} = \frac{n}{f} \frac{k_2^2}{k_6} [RH]^2 \qquad [34].\qquad (IV.1)$$

The same expression as we have seen above was obtained for non-catalytic reaction,* i.e. the catalyst does not increase the maximum rate but merely reduces the induction period. This seemingly paradoxical conclusion is based on the fact that in both cases the maximum rate is determined by the equality of catalyst-independent reaction rates of chain propagation which, in the maximum, are equal to the rate of chain branching and termination (reaction 6).

The existence of a maximum rate was quantitatively confirmed for the oxidation of some hydrocarbons (tetralin, ethylbenzene, diphenylmethane, etc.) [41, 42].

---

* If branching in non-catalytic reaction follows reaction 3, then the maximum rate of the catalytic process ($n/f = 3/4$) will be four times lower than that of the non-catalytic one ($n/f = 3$).

The maximum rate of dioxygen consumption

$$-\frac{d[O_2]}{dt} = k_{eff}[RH]^2$$

was found to be independent of catalyst concentration and the value k_{eff} to coincide quantitatively with the parameter $k_2^2/2k_6$ measured independently.

Chemiluminescent techniques have shown the hydroperoxide formation rate in these systems to be equal to that of dioxygen consumption. This is the evidence that the process proceeds entirely via the hydroperoxide formation [43].

Let us note that hydroperoxide decomposition in the presence of transition metal compounds may be in effect more complicated than follows from the above simple radical scheme. E.g., it may involve the formation of intermediate complexes between the catalyst and hydroperoxide. The reaction may result in the formation of molecular products without the participation of free radicals or proceed itself by the radical chain mechanism [44].

IV.2.3. CATALYTIC OXIDATION OF HYDROCARBONS IN THE PRESENCE OF LARGE CONCENTRATIONS OF TRANSITION METAL COMPOUNDS. CATALYST PARTICIPATION IN CHAIN PROPAGATION

The classical radical-chain mechanism leads to a variety of products in hydrocarbon oxidation due to the high temperature of the process and low selectivity of the interacting radicals. In view of developing selective industrial processes based on these reactions for the production of various substances, their low selectivity is an important shortcoming. The rate of oxidation usually increases in the order

$$RCH_3 < RCH_2OH < RCHO$$

Thus, oxidation products are oxidized in their turn faster than the hydrocarbons themselves, and therefore the yields of the required intermediate products are usually low.

As we have seen in many cases, the catalyst merely decreases the induction period, changing neither the mechanism nor the maximum rate of the process. This is typical of low catalyst concentrations, for example, in hydrocarbon media when metal salt concentration, due to low solubility, does not exceed several hundredths of one per cent. The use of polar solvents, e.g., acetic acid, trifluoroacetic acid, etc. allows a considerable increase (100–1000 times) of the catalyst concentration. In this case the reaction can acquire quite new features. Not only the rate but also the kinetics of the process can be markedly changed. Especially important is the fact that the process very often becomes highly selective, which allows it to be used on the industrial scale [45].

In a number of systems the reaction rate observed is several orders of magnitude higher than the maximum rate derived from equation IV.1.

Toluene oxidation by dioxygen at 60°C in acetic acid in the presence of Co(III) at the initial stages gives exclusively benzaldehyde [46]. The reaction proceeds without an induction period, reaching the maximum rate at the beginning of the reaction. The kinetic equation for oxidation is

$$-\frac{d[O_2]}{dt} = k\,[\text{Ph—CH}_3]\,[\text{Co(III)}]^2\,[\text{Co(II)}]^{-1}$$

It is noteworthy that the reaction rate increases sharply on the addition of sodium acetate, which may mean that the primary interaction of Co(III) with toluene, which the authors of [46] represented as electron transfer from toluene to Co(III), actually proceeds through electron transfer with synchronous elimination of the proton with the participation of a base (in this case, acetate).

The oxidation of alkylaromatic hydrocarbons proceeds particularly easily in the presence of both cobalt and bromide ions (a so-called 'cobalt-bromide catalysis'). Carboxylic acids are the final products of the reaction. For example, terephthalic acid is selectively formed from *p*-xylene, the whole process being used in the industrial production of the acid.

Despite the large number of works on cobalt-bromide catalysis, its mechanism has long remained speculative. The process has interesting kinetic characteristics; namely, in the presence of bromide ions the methylbenzenes oxidation rate is of second order with respect to cobalt. The rate does not depend on dioxygen concentration and depends only very insignificantly on the concentration of the compound being oxidized. To elucidate the 'cobalt-bromide catalysis' a variety of schemes has been suggested.

The reaction undoubtedly follows the chain mechanism (though some non-chain schemes were also suggested, some of them even without free radicals). The catalytic effect of cobalt and bromide ions is connected with their participation in chain propagation. Co(II) ions can interact with hydroperoxide radicals:

$$\text{RO}_2\cdot + \text{Co}^{2+} \longrightarrow \text{Co}^{3+} + \text{RO}_2^-$$

Cobalt(III) ions are again reduced to Co(II) by Br⁻ ions, the latter converting into bromine atoms [47, 48]:

$$\text{Co}^{3+} + \text{Br}^- \longrightarrow \text{Co}^{2+} + \text{Br}\cdot$$

Bromine atoms further react with hydrocarbons causing H elimination and forming a free radical

$$\text{Br}\cdot + \text{RH} \longrightarrow \text{HBr} + \text{R}\cdot$$

The bromide ions may enter the coordination sphere of Co(II) as ligands, facilitating the process of cobalt oxidation

$$RO_2 \cdot + Co(II)Br^- \longrightarrow Co(III)Br^- + RO_2^-$$

Then Br^- oxidation occurs in the coordination sphere of Co(III).

$$Co(III)Br^- \rightleftharpoons Co(II)Br \cdot$$

The reaction with hydrocarbons is suggested to proceed without $Br \cdot$ atoms leaving the coordination sphere of Co(II) [49]

$$Co(II)Br \cdot + RH \longrightarrow Co(II) + HBr + R \cdot$$

Table IV.3 gives the data on the relative rates of hydrocarbon oxidation obtained in the mixed oxidation of two hydrocarbons. (Usually the second hydrocarbon is ethylbenzene.)

TABLE IV.3

The relative reactivity of hydrocarbons in the reaction with active species and in autoxidation in the presence of Co^{2+} and Br^- in acetic acid

RH	Relative activity per one reactive H atom		
	Catalyst contains 0.02M of Co(OAc)$_2$ and 0.04M NaBr at 60°C	Cumyl-peroxy-radicals at 30°C	Br atoms at 40°C
Cyclohexane	0.33	0.1	0.0074
p-Chlorotoluene	0.42		
Toluene	1.00	1.00	1.00
Mesitylene	1.17		
p-Xylene	1.50	1.60	
Pseudocumene	2.54		
p-Methoxytoluene	3.42		
Durene	3.83		
Ethylbenzene	8.33	9.3	17
Cumene	16.8	15.9	37
Tetralin	34.2	36.4	

For comparison the relative reactivity for radicals and bromine atoms is also given (Table IV.3).

Kinetic studies combined with measurements of chemiluminescence resulting from the process of recombination of two radicals $RO_2 \cdot$ allow the proposal of a scheme where cobaltous, cobaltic and bromide ions enter the chain propagation [50, 51]:

(2′) $\qquad RO_2 \cdot + Co^{2+} + H^+ \underset{k_{-4}}{\overset{k_4}{\rightleftharpoons}} RO_2H + Co^{3+}$

(7) $\qquad Co^{3+} + Br^- \xrightarrow{k_7} Co^{2+} + Br \cdot$

(8) $\qquad Br \cdot + RH \xrightarrow{k_8} H^+ + Br^- + R \cdot$

The other stages are similar to those of the mechanism with participation of small catalytic concentrations

(1) $\qquad R \cdot + O_2 \longrightarrow RO_2 \cdot$

(2) $\qquad RO_2 \cdot + RH \xrightarrow{k_2} RO_2H + R \cdot$

(3″) $\qquad RO_2H + Co^{2+} \longrightarrow RO \cdot + Co^{3+} + OH^-$

(6) $\qquad 2 RO_2 \cdot \xrightarrow{k_6}$ products

Thus a slow reaction of chain propagation in the absence of the catalyst or when its concentration is small is replaced by more rapid reactions 2′, 7 and 8* with the catalyst participation. In this case the maximum rate is reached if

$$2k_{2'} [RO_2 \cdot] [Co^{2+}] = 2k_6 [RO_2 \cdot]^2$$

Hence, in agreement with the experimental results

$$v = - \frac{d[O_2]}{dt} = \frac{2k_{2'}^2}{k_6} [Co^{2+}]^2$$

Taking into account that hydroperoxide formation occurs in the reactions of $RO_2 \cdot$ not only with cobaltous ions but with hydrocarbons as well, we have, for steady-state conditions:

$$2k_2 [RO_2 \cdot] [RH] + 2k_{2'} [RO_2 \cdot] [Co^{2+}] = 2k_6 [RO_2 \cdot]^2$$

and

$$v = (2/k_6) (k_2 [RH] + k_{2'} [Co^{2+}])^2 \qquad\qquad (IV.2)$$

---

* In fact, reactions 2′, 7 and 8 are a catalytic pathway of the reaction 2.

Fig. IV.2 gives the relation of the square root of reaction rate and hydrocarbons concentration. In accordance with formula IV.2 the relation is linear, the slope of the straight line corresponding to the ratio $k_2(2/k_6)^{1/2}$.

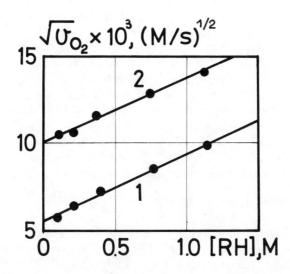

Fig. IV.2: The dependence of the square root of the oxidation rate on the concentration of hydrocarbons [50]. RH: ethylbenzene; solvent: acetic acid; [NaBr] = 8 × 10^{-3}M; [Co(OAc)$_2$] = 1 − 2.5 × 10^{-2}M, 2 − 50 × 10^{-2}M; t = 90°C.

The mechanism suggested is also supported by the fact that, in the case of alkylbenzenes, the dioxygen consumption rate coincides with that of the catalytic hydroperoxide decomposition [50, 51]. Moreover, the hydroperoxide radical recombination rate, measured by the chemiluminescence technique, is practically the same as that of dioxygen consumption (Table IV.4) [52].

TABLE IV.4

The dependence of dioxygen consumption rate, v_1, and peroxide radicals recombination rate, v_2, on cobalt(II) acetate concentration in toluene oxidation. [$C_6H_5CH_3$] = 1.9M; t = 80°C, [NaBr] : [Co(OAc)$_2$] = 0.1; solvent: acetic acid.

Co(OAc)$_2$ × 10^2 M	2.0	3.0	4.0	6.0	8.0
v_1 × 10^5 M s^{-1}	0.5	1.0	2.4	5.8	11.1
v_2 × 10^5 M s^{-1}	0.3	0.8	1.7	4.2	9.2

The bromide ion-catalyzed reaction of cobalt(III) with hydrocarbons is of special interest to us. In the suggested scheme Br^- catalyzes an electron transfer from RH to Co^{3+}

$$Co^{3+} + RH \xrightarrow{\ Br^- \ } Co^{2+} + R\cdot + H^+$$

The mechanism via bromine atoms is supported by molecular bromine formation in the interaction of Co^{3+} with Br^- in the absence of hydrocarbon (Br_2 is apparently formed by bromine atom recombination). This mechanism is also consistent with the fact that bromide ions, while catalyzing the oxidation in the case of alkylaromatic compounds, are not particularly effective in the case of simple alkanes. This corresponds to the difference of bromine atom reactivity with respect to alkylaromatic and aliphatic hydrocarbons. The bond energy in the H—Br molecule (85 kcal $mole^{-1}$) is practically equal to the energy of the C—H bond in the α-position to the aromatic ring, so that the reaction

$$Br\cdot + HR \longrightarrow HBr + R\cdot$$

for alkylaromatic hydrocarbons is close to thermoneutrality. For aliphatic hydrocarbons this reaction is endothermic and its activation energy is considerably higher, bromine ions ceasing to be an effective catalyst of electron transfer from RH to Co^{3+}.

The question arise as to why the reaction

$$Co^{3+} + RH \longrightarrow Co^{2+} + R\cdot + H^+$$

proceeds much slower without the catalyst, and why its rate is enhanced by Br^- ions. Probably the answer is that a direct electron transfer from RH to Co^{3+} requires a hydrocarbon molecule to approach Co^{3+} with H^+ being synchronously eliminated by the base. Apparently this is of rather low probability. At the same time, a negative bromine ion can easily enter the coordination sphere of a positive cobalt(III) ion, and the bromine atom produced will react with the hydrocarbon molecule without serious steric hindrance. It is obvious that such catalysis reflects the characteristics of the $Br\cdot/Br^-$ pair for which the redox potential for the optimum case must be close to those of the Co^{3+}/Co^{2+} and $(R\cdot + H^+)/RH$ pairs. As was mentioned above, Co^{3+} can rather effectively oxidize a number of hydrocarbons, even in the absence of bromide ions. Thus the following catalytic pathway is presumably realized

$$RO_2\cdot + Co^{2+} + H^+ \rightleftharpoons ROOH + Co^{3+}$$

$$Co^{3+} + RH \longrightarrow R\cdot + H^+ + Co^{2+}$$

in the presence of high cobalt salt concentrations and the absence of bromide. It looks especially interesting when applied to the alkanes, where the promoting action of bromide is weak. In effect, in some reactions of alkanes an unusual selectivity is observed in the presence of cobalt compounds, with or without dioxygen, which is strikingly different from that observed in radical reactions and which implicates the direct interaction of Co(III) with hydrocarbons.

Butane oxidation in the presence of cobalt(III) acetate in acetic acid occurs at temperatures of 100–125°C. Acetic acid is the reaction product with 83% selectivity (at 80% conversion) [53].

These data are markedly different from those observed for butane autoxidation at low initiator concentrations, where temperatures up to 170°C and higher are required and acetic acid is produced with 40% selectivity.

Cyclohexane oxidation in the presence of cobalt(II) acetate in acetic acid gives adipic acid in one stage as the main product with 75% selectivity at more than 80% cyclohexane conversion [54].

The induction period which is observed in the reaction decreases on the addition of promoters, such as acetaldehyde, or disappears completely if the cobalt salt is introduced in the three-valent state.

The kinetics of dioxygen consumption are described by the equation which is common for reactions with Co(III) participation

$$- \frac{d[O_2]}{dt} = k[RH] \, [Co^{III}]^2 \, [Co^{II}]^{-1}$$

It is noteworthy that benzene under these conditions is not oxidized and is an appropriate solvent for cyclohexane oxidation. However, in a mixture of benzene–acetic acid the rate is 40% higher than that in acetic acid alone.

The alkanes containing a tertiary C—H bond appear to be less reactive than normal alkanes (the same is observed in the absence of dioxygen), the reactivity surprisingly decreasing with the increase of the hydrocarbon chain length. Thus, under similar conditions isobutane reacts slower than butane, n-butane is more reactive than n-pentane, undecane being completely unreactive [53]. It is difficult, as yet, to find any plausible explanation for these results.

Heptane oxidation by cobalt(III) acetate in trifluoroacetic acid proceeds selectively without dioxygen in position 2 (see Section IV.1.3), and in the presence of dioxygen results in 2-heptanone with 83% selectivity [18]. Undoubtedly, the alkane in these reactions interacts with cobalt(III) salt and the selectivity observed, as has already been pointed out, is of special interest, indicating the significant role of steric factors.

The kinetics of the reaction, however, seem to be complicated by the oxidation of substances produced from alkanes. These secondary reactions might proceed according to a mechanism different from the mechanism of alkanes oxidation.

IV.3. Biological Oxidation of Alkanes

Versatile oxidation reactions play an immense role in all organisms. Enzymes participating in these reactions usually involve metal ions; those of iron, copper, cobalt, manganese, molybdenum, and vanadium. Substrates subjected to oxidation include molecules with activated and non-activated C—H bonds. Among them are saturated hydrocarbons including methane and its homologs.

Enzymatic reactions are widely known to possess outstanding characteristics, unachievable as yet in 'traditional' chemistry; those of high rates, selectivity, and stereospecificity. Oxidation processes such as, for example, hydroxylation of C—H containing compounds, including chemically inert alkanes, proceed with high rates at low temperatures.

Elucidating the mechanism of these reactions might help to find similar purely chemical processes which could be of particular interest both as models of biological processes and as new ways of transforming hydrocarbons.

The development of the knowledge of biological oxidation mechanisms has already had a long history [55]. As early as the beginning of the 20th century Bach expanded his ideas of the mechanism of chemical oxidation to cover the field of biological oxidation, assuming the formation of intermediate peroxide compounds with the participation of enzyme oxygenase [56].

$$A + O_2 \xrightarrow{\text{oxygenase}} A \diagdown \begin{matrix} O \\ | \\ O \end{matrix}$$

Peroxides further oxidize substrates in peroxidase-catalyzed reactions

$$A \diagup \begin{matrix} O \\ | \\ O \end{matrix} + X \xrightarrow{\text{peroxidase}} AO + XO$$

Later on Warburg put these ideas into more concrete form, assuming that iron ions involved in the heme molecule take part in dioxygen activation [57]

$$Enz-(Fe) + O_2 \longrightarrow Enz-(Fe)-O_2$$

$$Enz-(Fe)-O_2 + 2X \longrightarrow Enz-(Fe) + 2XO$$

There is much evidence, however, that the participation of dioxygen in many cases of enzymatic oxidation is not necessary and O_2 can be replaced by other oxidants, such as methylene blue. Taking this into account, Wieland developed a scheme of biological oxidation [58] where dioxygen did not participate directly, but where substrate molecule XH_2 transferred electrons to a corresponding acceptor A:

$$XH_2 + A \xrightarrow[\text{reduction}]{\text{oxidation}} X + AH_2$$

For a large number of cases this scheme was confirmed and became widely accepted. The enzymes catalyzing these reactions were named dehydrogenases. Usually, pyridine-nucleotides, flavin-nucleotides and cytochromes are electron acceptors in biological oxidation. In the cases of enzymatic processes where O_2 is itself a direct electron acceptor, the enzymes are named oxidases. Later on, enzymes were discovered which catalyzed the reaction, not only in the obligatory presence of dioxygen, but where oxygen atoms from O_2 entered into the molecule produced from the substrate. Such enzymes were named oxygenases. At present several types of oxidases and oxygenases are known [55].

The first group of oxidases involves enzymes catalyzing electron transfer to dioxygen to give a superoxide anion.

$$XH + O_2 \longrightarrow X + O_2^- + H^+$$

The second group catalyzes a two-electron transfer with formation of hydrogen peroxide:

$$2 XH + O_2 \longrightarrow 2 X + H_2O_2$$

The third group includes oxidases catalyzing four-electron transfer to dioxygen to give two water molecules:

$$4 XH + O_2 \longrightarrow 4 X + 2 H_2O$$

(cytochrome oxidase, ascorbate oxidase, ceruloplasmin are the examples of this group of oxidases). In this case, several metal ions are present, as a rule, in the enzyme molecule.

Oxygenases are divided into two types, depending on whether one or two atoms of the dioxygen molecule enter the product molecules. They are named, respectively, dioxygenases and monooxygenases.

Intramolecular dioxygenases catalyze reactions where both oxygen atoms enter the molecule of the product. Intermolecular dioxygenases are enzymes catalyzing the introduction of two oxygen atoms in two different substrates. One of these substrates is practically always α-ketoglutarate which is **oxidized** to succinate, this oxidation being coupled with the insertion of an oxygen atom into the molecule of another substrate.

Internal monooxygenase catalyzes the insertion of one oxygen atom of the O_2 molecule into the substrate molecule, using its electrons to form water from another oxygen atom according to the scheme

$$XH_2 + O_2 \longrightarrow XO + H_2O$$

External monooxygenases, which are a more widespread type, include enzymes requiring the participation of various electron donors. In this case the oxidation of a substrate is coupled with the oxidation of the donor molecule via the scheme

$$X + O_2 + AH_2 \longrightarrow XO + A + H_2O$$

They are, for example, heme-containing monooxygenases oxidizing the substrate molecule together with electron donors such as reduced nicotinamide nucleotide (NADH) or flavin adeninedinucleotide ($FADH_2$).

As was already indicated, the reactions of biological oxidation in most cases require one or several transition metal ions to participate in the enzyme molecule, and these may change oxidation state in the course of the reaction. When the aliphatic C—H bond in a substrate is oxidized, the participation of a metal ion seems to be always required. Let us now review three groups of such enzymes.

IV.3.1. α-KETOGLUTARATE-COUPLED DIOXYGENASES [59]

α-Ketoglutarate-coupled dioxygenases catalyze various reactions of biological oxidation; in particular, the hydroxylation of chemically non-activated C—H bonds. The reaction with the substrate proceeds via the following general scheme:

$$
O_2 + X + \underset{\substack{\displaystyle| \\ \displaystyle CH_2 \\ \displaystyle| \\ \displaystyle CH_2 \\ \displaystyle| \\ \displaystyle COCOOH}}{\overset{\displaystyle COOH}{}} \quad \xrightarrow[\text{(Fe}^{2+}\text{)}]{\text{reductant}} \quad XO + \underset{\substack{\displaystyle| \\ \displaystyle CH_2 \\ \displaystyle| \\ \displaystyle CH_2 \\ \displaystyle| \\ \displaystyle COOH}}{\overset{\displaystyle COOH}{}} + CO_2
$$

α-ketoglutarate   succinate

The cases presented below of α-ketoglutarate-coupled dioxygenase hydroxylation can serve as examples of hydroxylation of the aliphatic C—H bond (Table IV.5).

TABLE IV.5
Examples of α-ketoglutarate-coupled dioxygenase hydroxylation

Enzyme	X	XO
Prolyl hydroxylase	(proline residue with H on ring, $C=O$ and NHR, N–C=O–R')	(hydroxyproline residue with HO on ring, $C=O$ and NHR, N–C=O–R')
Lysyl hydroxylase	CH_2NH_2 — CH_2 — $(CH_2)_2$ — $R-NH-CH-COR'$	CH_2NH_2 — $CHOH$ — $(CH_2)_2$ — $R-NH-CH-COR'$
γ-Butyrobetaine hydroxylase	$(CH_3)_3 \overset{+}{N}-CH_2-CH_2CH_2COO^-$ γ-butyrobetaine	$(CH_3)_3 \overset{+}{N}-CH_2-CHOH-CH_2COO^-$ carnitine
Thymine-7-hydroxylase	(thymine ring with CH_3) thymine	(uracil ring with CH_2OH) 5-hydroxymethyluracil

In all the cases the reaction requires the presence of iron ions which are highly specific. Any attempts to replace iron in purified enzymes by other metal ions (such as Mg^{2+}, Mn^{2+}, Cu^{2+}, Cd^{2+}, etc.) inevitably lead to a complete loss of catalytic activity.

Experiments with labelled dioxygen $^{18}O_2$ have proved that one oxygen atom of the O_2 molecule is inserted into the XO product, and the other into the succinate molecule formed from α-ketoglutarate.

The mechanism of oxidation is not well-studied yet, although it might be close to that of monooxygenase action in the case of cytochrome P-450 described below.

IV.3.2. CYTOCHROME P-450

The biological oxidation of various organic substances including those with aliphatic C—H bonds has been studied in detail for cytochrome P-450 containing monooxygenases (see, for example, [60]).

These monooxygenases function in a variety of living organisms ranging from bacteria to higher mammals, including man. Cytochrome P-450 is a hemoprotein named after an abnormal position in the absorption spectrum of the reduced complex with carbon monoxide (at 450 nm).

It has been found in tissues of various animals: in liver, lung, kidney, cutis. Monooxygenases connected with cyctochrome P-450 functioning in liver microsomes have been extensively studied. Their functions are related to the oxidation of various foreign compounds via oxygen atom insertion into products, being subsequently removed from the organism. The substrates include aliphatic and aromatic hydrocarbons, amines, alcohols, phenols, thiophenols, as well as steroids and fatty acids which are harmful to the organism. The reaction can proceed as hydroxylation, oxidative dealkylation or deamination, formation of aminooxides or sulfoxides.

Monooxygenases of aerobic bacteria can convert inert molecules by oxidation into more chemically active ones capable of undergoing further enzymatic interactions.

Actually, there exists a whole set of monooxygenases. For example, the liver of any organism involves several quite different enzymes united by the use of cytochrome P-450. Evidently, they can catalyze oxidation reactions of various substrates, i.e. possess different substrate specificity.

From the chemist's point of view, the most striking is the selectivity at a site of attack when oxidizing C—H-containing compounds. Thus, in the oxidation of alkanes and fatty acids in the presence of mammalian liver microsomes, the hydroxylation of a terminal (so-called ω) carbon atom occurs initially, i.e. the most stable C—H bond of the methyl group is attacked [61]. Besides the terminal C—H bond the neighbouring secondary C—H bond ($\omega - 1$) often undergoes hydroxylation as well. Thus about 92% of 10-hydroxydecanoic and 8% of 9-hydroxydecanoic acid are produced in the hydroxylation of decanoic acid in microsomes of rat's liver. The ω- and ($\omega - 1$)-hydroxylation might be catalyzed by different monooxygenases, since carbon monoxide, which is an inhibitor of oxidation catalyzed by cytochrome P-450, retards the ($\omega - 1$)-hydroxylation less strongly than ω-hydroxylation. ($\omega - 1$)-Hydroxylation of 9-$^2H_2$ decanoate occurs with a significant isotope effect in contrast to ω-hydroxylation (of 10-$^2H_3$-decanoate) where this effect is not observed. For example, for ($\omega - 1$)-hydroxylation of lauric acid labelled by deuterium in position 11 [11-$^2H_2$-lauric acid] the maximum reaction rate is 3.6 times lower than for hydroxylation of non-labelled acid [62].

Fatty acids and alkanes in yeast and bacterial systems also undergo ω- and $(\omega - 1)$-hydroxylation. The relation of ω- and $(\omega - 1)$-hydroxylation often depends on chain length and degree of unsaturation of fatty acids.

It is essential that hydroxylation occurs stereospecifically. Thus mainly L-isomers of 9-hydroxydodecanoic acid are formed in the above mentioned $(\omega - 1)$-hydroxylation of lauric acid.

When using $[9L-^2H]$ decanoic acid, the hydroxylation in the presence of cytochrome P-450 was shown to occur with retention of the optical configuration.

It should be noted that enzymatic oxidative dehydrogenation of fatty acids which occurs in the presence of dioxygen is also stereospecific. This was checked by converting stearate into oleate, which was performed by *Corynebacterium diphteriae* microorganism. Using stereospecifically tritium-labelled stearate it was shown that hydrogen atoms which were removed from the 9- and 10-positions corresponded to a D-configuration in the initial stearate.

Besides biological systems displaying a selectivity unusual in chemistry (ω- and $(\omega - 1)$-hydroxylation) there exists non-specific monooxygenase extracted from mammalian liver microsomes (e.g., from the rat) where the 'normal' activity of the C—H bond hydroxylation is observed: $3° > 2° > 1°$ [63, 64].

For example, the ratio of hydroxylation rates of secondary and primary C—H bonds $(RS_{s/p})$ in 2-methylbutane (taking into account the number of corresponding C—H bonds in the molecule) is 15.4 ± 4.2, whereas for tertiary and secondary bonds, $RS_{t/s}$ is 7.3 ± 0.5. For methylcyclohexane, $RS_{s/p}$ = 24 ± 14 and $RS_{t/s}$ = 3.0 ± 0.3. $RS_{s/p} > 50$ for n-pentane and the ratio of reactivities of C-2 and C-3 atoms in hydrocarbons is 2.6 ± 0.9.

As compared with chemical systems (see Section IV.1), a somewhat lower reactivity is observed for the tertiary C—H bond and a somewhat higher reactivity (2—3 times) is observed for the secondary C—H bond of a second (C-2) pentane carbon atom than of a third one (C-3). However, this difference in selectivity becomes apparent only with quantitative estimation.

Cyclohexane is hydroxylated by non-specific monooxygenase without any isotope effect ($k_H/k_D \simeq 1$).

2,2-Dimethylpropane, which contains only primary C—H bonds, is hydroxylated at virtually the same rate as the other alkanes (n-butane, 2-methylbutane, etc.) including secondary and tertiary C—H bonds [63, 64], ethane not being hydroxylated at all. Apparently cytochrome P-450 can form a reversible enzyme-substrate complex with hydrocarbon, while the rate determining step is the reduction of this complex or the reduction of an Fe(II)-substrate-O_2 complex.

The further formation of an active form of the complex in the reaction with O_2 and the substrate hydroxylation occur in more rapid stages and the differences in substrate reactivities do not affect their hydroxylation rates. However, when different bonds are present in the same molecule, the different reactivities of the bonds will naturally influence the ratio of the products. Small hydrocarbon molecules, such as ethane, do not apparently form the enzyme-substrate complex and, hence, are not hydroxylated.

Based on hexane, it was demonstrated [65] that cytochrome P-450-linked monooxygenase extracted from rat liver preferentially hydroxylates the hydrocarbon in position 2 (($\omega - 1$)-hydroxylation). Initiators and inhibitors affect the hydroxylation in different ways in positions 1,2, and 3, indicating the presence of more than one monooxygenase, even in the same mammalian organ.

In the hydroxylation of aromatic compounds, monooxygenase usually behaves as an electrophilic reagent with preferentially *ortho-para*-orientation in the case of hydroxylation of substituted benzenes with electron-donating substituents.

For example, in acetanilide hydroxylation the isomer ratio $m/p \approx 0.01$ in products (hydroxyacetoanilides); however, a higher m/p ratio is observed in cresol formation in hydroxylation of toluene ($m/p = 0.47$) [66].

A so-called NIH-shift* [67] is observed in hydroxylation of aromatic compounds in the presence of cytochrome P-450. It consists in the shift of the hydrogen atom bound to the hydroxylating carbon to the neighbouring position along with the insertion of the oxygen atom into the C—H bond to produce phenol. It was discovered in the study of deuterium-labelled aromatic molecules:

$$\text{X-C}_6\text{H}_4\text{-D} + [\text{O}] \xrightarrow[\text{monooxygenase}]{\text{Cyt. P-450}} \text{X-C}_6\text{H}_3(\text{D})(\text{OH})$$

The efforts of many investigators directed at elucidating the nature of cytochrome P-450-linked monooxygenase action have now succeeded in establishing a rather definite picture of its structure and mechanism of action.

Cytochrome P-450-linked monooxygenase may be divided into several components. Thus, three fractions involving cytochrome P-450 were isolated for microsome monooxygenase, those containing cytochrome P-450, NADPH-cytochrome-*c*-reductase, and 'heat stable factor'. Reductase was shown to

---

* NIH = National Institute of Health — the Institution where the effect was discovered.

catalyze an electron transfer from NADPH to cytochrome P-450 under anaerobic conditions. The 'heat stable factor' is apparently a phospholipid and may be replaced by phosphatidyl choline, the additions of the latter to the other two components leading to the restoration of enzymatic activity. It is interesting to note that the efficiency of phosphatidyl choline depends on the nature of the derivatives of fatty acids involved in its structure. The highest activity is displayed by molecules with oleoyl residues and one or two lauroyl residues.

The experimental evidence testifies that phospholipid is necessary as a structural factor for enzymatic reduction of cytochrome P-450 [68].

The bacterial enzymatic systems involve somewhat different components. For example, monooxygenase in *Pseudomonas oleovorans* involves three protein components, those of rubredoxin, NADH-rubredoxin reductase and ω-hydroxylase. Evidently, two rubredoxine-containing iron atoms are an electron-transfering agent from NADH to the enzyme active centre.

Therefore, various monooxygenases can include various electron donors and electron-transfer agents from the donor to the active centre. The principal mechanism of various monooxygenases, however, remains the same. The following scheme (Scheme IV.3.2.1) visualizes the functioning of a microsomal hydroxylating system [69]:

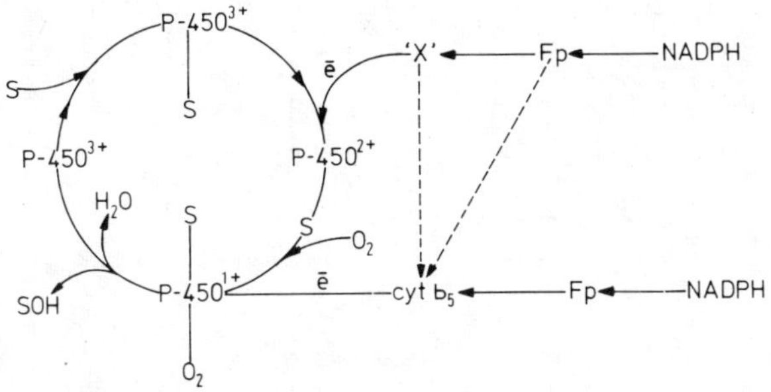

Scheme IV.3.2.1.

The substrate molecule forms a complex with the oxidized state of cytochrome P-450. Then NADPH transfers one electron to the enzyme-substrate complex with the help of a corresponding transfer agent. NADH can also serve as an electron donor (although less effective).

The next step of the process is the addition of the dioxygen molecule to the reduced cytochrome to produce the complex. The latter becomes active with respect to the substrate only after the transfer of the second electron from the reducing agent. Subsequently the substrate hydroxylation takes place to produce water molecules, which is again followed by a new substrate molecule joining the oxidized state of cytochrome P-450.

Thus, as has been mentioned before, the substrate oxidation occurs in a coupled reaction with the oxidation of a biological reducing agent, in this case, NADPH.

Naturally, most interesting, from the chemist's point of view, is the detailed structure of the active centre of cytochrome P-450 and the mechanism of its reaction with a substrate.

Similarly to other prophyrin metal complexes, four coordination sites in the iron ion involved in the active centre of cytochrome P-450 are coupled by N atoms from the porphyrin molecule (Fig. IV.3). The fifth position is

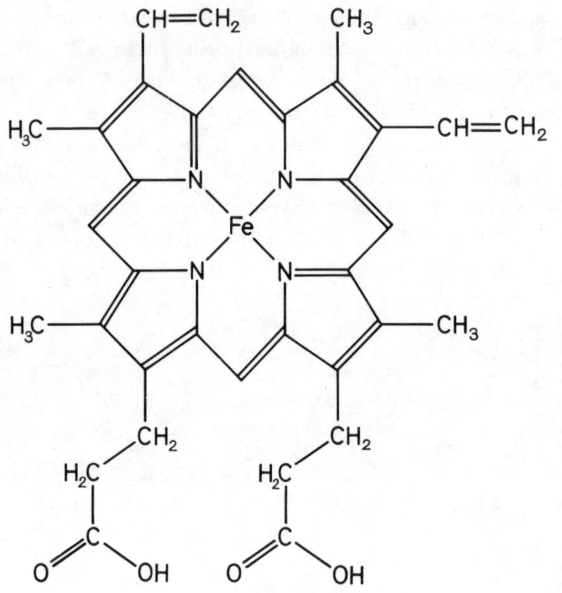

Fig. IV.3: The prosthetic group of cytochrome P-450: heme (iron-protoporphyrin IX).

occupied by a sulfur atom from the cysteine molecule, which distinguishes it from hemoproteins involved in other enzymes, such as catalases and peroxidases (where the fifth position is occupied by a nitrogen atom of the histidine molecule). In the absence of a substrate the sixth coordination place is probably occupied by histidine or more probably OH^- [70]. The ESR spectrum of this oxidized state of cytochrome P-450 with g-factor values: $g_z = 2.4$, $g_y = 2.2$, and $g_x = 1.9$ reveals a low-spin iron state in the complex which is converted to a high spin state after adding the substrate. The introduction of the substrate changes the complex into a five-coordinated one, liberating the axial ligand trans to the cysteine. Taking into consideration the chemical nature of cytochrome P-450 substrates, in particular that of alkanes, it is unlikely that they are chemically bound to the enzyme in the enzyme-substrate complex. Instead, some data show the existence of a so-called hydrophobic pocket in the close vicinity of the enzyme active centre. Thus the substrate hydrophobic interaction may keep it bound to the enzyme [71, 72].

The formation of the enzyme-substrate complex is the triggering event stimulating iron reduction. The reason is possibly the configuration change and/or increase of redox potential under the action of the susbtrate molecule.

The subsequent dioxygen addition is followed by the active complex formation (which requires one more electron transfer) and then by the reaction with the substrate. The mechanism of an elementary process of enzymatic hydroxylation has been discussed in a number of papers, different mechanisms being suggested. The problem cannot be solved using the existing chemical models, since they function via various mechanisms (see below). Since the process in many aspects parallels the reactions with carbenes (addition to the multiple bond, insertion into the C—H bond), Hamilton suggested a so-called oxenoid mechanism with an oxygen atom ('oxene') inserted into the C—H bond without an intermediate formation of free radicals [73, 74]. The source of oxygen atoms might be a complex of the oxygen molecule with cytochrome Fe ion:

$$\text{Fe} \cdots \text{O} \dot{-} \text{O} + \begin{array}{c} \text{R} \\ | \\ \text{H} \end{array} \longrightarrow \text{Fe}=\text{O} + \text{ROH}$$

Alongside with this proposal there are suggestions [71] of a possible free-radical mechanism of H atom abstraction with a subsequent reaction without the radical leaving the 'cell' where the reaction occurs:

$$\text{Fe}^{2+}-\text{O}\dot{-}\text{O}^- + \text{RH} \longrightarrow \text{R}\cdot + [\text{Fe}-\text{O}-\text{OH}]^+ \longrightarrow [\text{FeO}]^+ + \text{ROH}$$

Recently, a mechanism has been put forward postulating that in the active state of cytochrome P-450 only one oxygen atom is bound to iron in the form of a ferryl ion species FeO^{3+} [75, 76].

The oxidation in the presence of cytochrome P-450 was found to occur without dioxygen with the participation of single oxygen donors, hydroperoxides, peroxy acids, and iodosobenzene [77]. This result, especially for iodosobenzene suggests that FeO^{3+} might be the source of 'oxene', ferryl being the result of the proton's action upon the O_2 complex with iron

$$[\text{Fe} \cdots \text{O}_2]^+ + 2\text{H}^+ \longrightarrow [\text{FeO}]^{3+} + \text{H}_2\text{O}$$

The electronic structure of this complex can be presented by the following forms [70]:

$$-S^- \cdots \overset{\backslash\,/}{\underset{/\,\backslash}{Fe}}{}^{IV}-\overline{\underline{O}}\cdot \quad \text{or} \quad -S^- \cdots \overset{\backslash\,/}{\underset{/\,\backslash}{Fe}}{}^{V}=O$$

$$-S-\overset{\backslash\,/}{\underset{/\,\backslash}{Fe}}{}^{III}-O\cdot \quad \text{or} \quad -S-\overset{\backslash\,/}{\underset{/\,\backslash}{Fe}}{}^{IV}=O$$

The participation of a porphyrin ring in the oxidized state of the complex can be visualized with electron transfer to iron forming a cation-radical form of the porphyrin. Dolphin *et al.* [78, 79] showed that the absorption spectra of complex-I of peroxidase and catalase are similar to those of metal porphyrin complexes with a π-cation radical of a porphyrin ring. Taking this into account the oxidation state of Fe in the forms above will be then one unit less. The electronic structure of the oxidized complex parallels that of a so-called peroxidase-I compound. According to the oxenoid mechanism the complex produced must be capable of transferring the oxygen atom to insert it into the C—H bond or to add to the unsaturated compounds.

Although the final results of hydroxylation with the participation of mono-oxygenases is the transfer of the oxygen atom from O_2 molecule into the water molecule, this O atom can be primarily bound to an organic acylating group. Such an acylating group is evidently required by bacterial camphorhydroxylase (unlike P-450 proteins of highest mammals) in order to be able to function in the catalytic cycle. In native hydroxylase this group is complex-bound with P-450 hemoprotein due to the presence of carboxy-terminal tryptophan or of the penultimate glutamine putido-redoxine. Purified P-450 from *Pseudomonas* cannot hydroxylate camphor in the absence of the effector molecules, such as, e.g., lipoic acid. The formation of superoxide occurs in the absence of the effector instead of the expected hydroxylation product. It was shown in [80], using $^{18}O_2$ and dihydrolipoic (2,3-tioctic) acid, that at the initial stages one ^{18}O atom is included in the carboxyl group for each ^{18}O atom in the hydroxylation product (*exo*-5-hydroxycamphor). Therefore, water is formed from the initially unlabelled hydroxy group of dihydrolipoic acid.

Taking this result into account two different schemes (Schemes IV.3.2.2 and 3) may be suggested and the data of the distribution of ^{18}O atoms are in agreement with the assumption of peroxide formation as an intermediate product [74] (Mechanism A).

Assuming a heterolytic mechanism with the intermediate formation of $[FeO]^{3+}$, it should be concluded that OH^- elimination from the dihydrolipoic acid molecule proceeds before the O—O bond splitting. The similarity in the mechanism of action of peroxidases and monooxygenases should be noted, which is illustrated by the scheme [80].

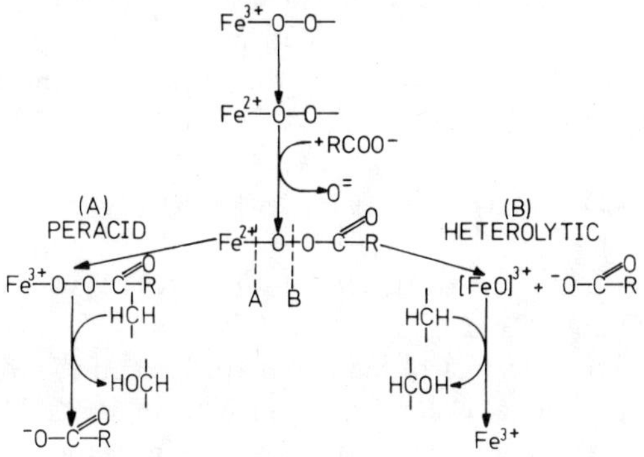

Scheme IV.3.2.2

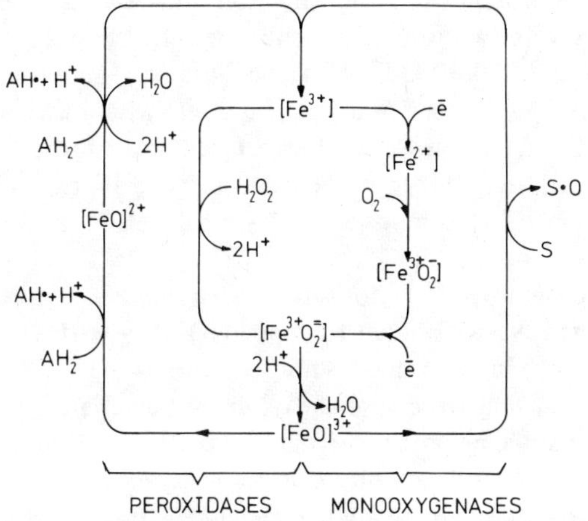

Scheme IV.3.2.3

IV.3.3. METHANEMONOOXYGENASE

Enzymes which can catalyze the oxidation of methane, the simplest and the most chemically inert saturated hydrocarbon, are particularly interesting (see reviews [71, 81, 82]).

These enzymes are involved in various methane-oxidizing bacteria, numbering about a hundred. The first data on the existence of methane oxidizing bacteria in nature were reported more than 70 years ago. Nevertheless, at the present time, only four varieties of them have been thoroughly investigated and described. Cell-free preparations for oxidizing methane were obtained only quite recently. Hence, the first data on the chemistry of oxidation appeared in the literature within the last few years. Methane oxidation in the presence of methanemonooxygenase is described by the general scheme characteristic of monooxygenases

$$CH_4 + O_2 + AH_2 \longrightarrow CH_3OH + H_2O + A$$

where AH_2 is the donor of two electrons (usually NADH which can be replaced by NADPH or sodium ascorbate).

Methanemonooxygenase is isolated and partially purified from two different methane oxidizing bacteria. It consists of three components. Similarly to other monooxygenases, there must be a protein component taking part in reducing an active component (reductase), an electron-transport protein, and methanemonooxygenase itself. The composition of enzymes from *Methylococcus capsulatus* (Bath) consists of a protein with non-heme iron (m.w. *ca*. 220 000), flavoprotein with a centre involving two iron atoms and two sulfur atoms (m.w. *ca*. 44 000) and protein (m.w. *ca*. 15 000) [81]. Methanemonooxygenase from *Methylosinus trichosporium* involves copper-containing protein with m.w. 47 000, soluble CO-binding cytochrome-*C* (m.w. *ca*. 13 000) and small protein with m.w. *ca*. 9400. By ESR analysis, copper of a so-called 'type II' (see [71]) was identified in cytochrome-C_{CO} composition. Based on NMR spectroscopy data this cytochrome was supposed to bind substrates [83].

When studying membrane fractions from *M. capsulatus*, an intense ESR signal was recorded which was attributed to copper(II) of type II. However, the role of the copper in the monooxygenase is not yet quite clear.

Cytochrome-*C* isolated from methane oxidizing bacteria of *M. capsulatus* also contains copper, producing an ESR signal of type II, and it does not differ from cytochrome-C_{CO} of *M. trichosporium* in its general characteristics. However, the addition of this cytochrome to the membrane fraction does not result in an increase of this methanemonooxygenase activity. Moreover, with cytochrome-C_{CO} completely removed from the membrane fraction, the activity of methanemonooxygenase left in the membrane does not change. Therefore, cytochrome-C_{CO} might not necessarily be a substrate-binding protein. It is also possible that reducing and electron-transport components of methanemonooxygenase in various microorganisms are different. At present, the nature of a substrate-binding protein component still remains unclear.

The study of the kinetics and mechanism of methanemonooxygenase action has recently helped to gain more knowledge of this subject.

The optimum pH value for monooxygenases from various microorganisms lies in the narrow range of pH about 7.0 [84, 85].

Cell-free preparations of methane oxidizng bacteria can oxidize other n-alkanes up to C_8, as well as alkenes, cyclic and aromatic hydrocarbons, and carbon monoxide [86]. The latter yields CO_2 when oxidized. In the case of $M.$ $capsulatus$, alkanes initially produce alcohols, n-alkanes giving exclusively 1- and 2-alcohols with no 3- and 4-alcohols. Alkenes are oxidized to epoxides. cis- and $trans$-but-2-enes are both oxidized to give 2,3-epoxybutanes and but-2-ene-1-ol, cis- and $trans$-configuration being preserved in both the cases; cis-but-2-ene can also give 2-butanone. Cyclohexane is hydroxylated into cyclohexanol, this reaction proceeding more readily than hexane oxidation. Hydroxylation of benzene produces phenol. Toluene is oxidized yielding both cresol and benzyl alcohol. Styrene gives the epoxide and p-hydroxystyrene. Pyridines are selectively oxidized into N-oxides.

These results, as well as the analogous results for cytochrome P-450, may evidently be explained by a combination of steric and electronic factors: the CH_3 group is the most open but its binding energy is higher than of the CH_2 group which, in position 2, is still accessible for the reaction to proceed. In positions 3, 4, etc. these groups are unreactive due to steric hindrances.

The comparison of oxidation rates of methane and deuteromethane CD_4 shows the reaction to have a considerable isotope effect (k_{CH_4}/k_{CD_4} = 5 ± 0.2) which implies that the C—H bond cleavage is involved in the rate determining step [87].

A specific rate of methane oxidation is 20—50 times higher than that of saturated hydrocarbons catalyzed by cytochrome P-450.

The mechanism of binding methane and its homologs in an enzyme-substrate complex is of considerable interest. As has been pointed out for cytochrome P-450, the complex is formed by a so-called hydrophobic interaction. The latter cannot be significant for the small methane molecule. Carbon monoxide, ethylene, and acetylene — which are presumably chemically bonded in the enzyme-substrate complex — are the competitors for monooxygenase reaction. Hence, the assumption is made that methane and its homologs are chemically activated upon entering into the coordination sphere of a metal (iron or copper) [71].

According to this hypothesis, alkane and oxygen molecules are activated by two neighbouring ions, followed by the reaction of alcohol formation.

IV.3.4. ON THE MECHANISM OF BIOLOGICAL OXIDATION

Three groups of enzymes oxidizing the non-activated C—H bond, which were considered above, in spite of some differences, possess important common

features, making it possible to suggest a general mechanism of C—H hydroxylation. Indeed, all the enzymatic reactions with non-activated C—H bond are the
processes of coupled oxidation of substrate and electron donors (NADPH,
NADH, ascorbate). There is no doubt that in the case of ketoglutarate-dependent
dioxygenase and cytochrome P-450, iron is involved in the active centre and iron
may also be a participant in the methanemonooxygenase active centre.

The role of ketoglutarate in the case of ketoglutarate-dependent dioxygenase
may consist in accepting the OH^- ion during the active state formation:

$$[FeO_2]^+ + \quad \begin{array}{c} | \\ C=O \\ | \\ C \\ HO \diagup \quad \diagdown O \end{array} \quad \xrightarrow{H^+} \quad FeO\!-\!O \quad \begin{array}{c} H \\ | \\ \overset{|}{O} \\ \end{array} \quad \begin{array}{c} | \\ C=O \\ | \\ C \\ HO \diagup \quad \diagdown O \end{array} \longrightarrow$$

$$[FeO]^{3+} + HO\!-\!\overset{|}{\underset{\underset{HO \diagup \ \diagdown O}{C}}{C}}\!-\!O^-$$

$$RH \diagup \qquad (Fe^{3+}) \qquad \diagdown \ -2e$$

$$ROH + Fe^{3+} \qquad\qquad \overset{|}{\underset{O \diagup \ \diagdown OH}{C}} \quad + CO_2 + H^+$$

As was shown for cytochrome P-450, at least in the case of *Pseudomonas*, an
organic acylating group is also used for oxygen activation, probably via OH^-
elimination [80].

An analogous active centre may be supposed in the case of methanemonooxygenase, though a non-heme iron complex (possibly a dimer) is apparently
active here. Actually, methanemonooxygenase and monooxygenase with the
participation of P-450 have more common features than differences. Both the
enzymes catalyze the epoxydation of olefins and hydroxylation of aromatic and
aliphatic hydrocarbons, methyl and neighbouring methylene group hydroxylating
for normal hydrocarbons of the homologous methane series (ω- and ($\omega - 1$)-
hydroxylation). In both the cases the C—H bond hydroxylation may proceed
with a considerable isotope effect (if oxidation is the rate determining step).

The differences are mainly connected with different activities in the hydrocarbon series: the maximum activity for cytochrome P-450 is observed for C_8-C_{12} hydrocarbons, and for methanemonooxygenase methane is the most active, the ability of enzyme to oxidize the other alkanes decreasing rapidly with an increase of molecule sizes. These differences can be explained by variations in the sizes of the cavity adjacent to the active centre, rather than the ultimate changes of the mechanism.

Therefore, the data reported up to now allow the presentation of the cases of enzymatic hydroxylation of the aliphatic C—H bond as a mechanism with a high-valent iron form, FeO^{3+}.

The reaction with the non-activated C—H bond apparently requires a high value of redox potential of the active species and it is easier to visualize it with $[FeO]^{3+}$ than with any form of peroxocomplex (see Section IV.5). The surroundings of the iron atom must fulfil two tasks, the one is to facilitate OH^- abstraction from $[FeOOH]^{2+}$ and the other is to keep a high value of redox potential of the formed $[FeO]^{3+}$ species. These tasks seem to require opposite influences of ligands attached to the active centre. Therefore, some optimum has to be achieved. In this way we can explain the necessity of the neighbouring S atom in the case of cytochrome P-450 (facilitating OH^- abstraction). In the case of methanemonooxygenase, another iron atom situated close to the active centre may also facilitate the OH^- abstraction to form $Fe^{IV} \cdots Fe^{IV}=O$ instead of $Fe^V=O$).

Nature, which is known to be very economical, need not invent a new mechanism for methane oxidation different in principle from that used for cytochrome P-450. In effect, even for non-specific monooxygenase from liver, where tertiary C—H bonds are preferred, the reaction with neopentane, having only primary C—H bonds of methyl groups, proceeds readily.

Although this conclusion is not unequivocal and might be disproved later, we can use the same mechanism for considering the results of biological hydroxylation of the aliphatic C—H bond, particularly in the case of non-activated C—H bond in CH_4 and CH_3 groups.

We shall discuss the mechanism of enzymatic hydroxylation of alkanes along with the other cases of the interaction of electrophilic oxidants with alkanes at the end of this chapter.

IV.4. Coupled Oxidation of Hydrocarbons and Metal Complexes. Chemical Models of Monooxygenases

The biological oxidation of hydrocarbons catalyzed by monooxygenases is coupled with the oxidation of electron donors, such as NADH or NADPH in the presence of iron complexes. Hence, it is only natural to search for chemical models of these enzymatic reactions among coupled oxidations with the participation of metal ions. The donor in biological oxidation is believed to transfer its electrons initially to the metal ion, which is subsequently oxidized by an oxygen molecule. Therefore, it is possible first to set up the simpler task of studying the possibility of coupled oxidation of hydrocarbons and metal or ion complexes in a low-valent state. It turns out that this problem can be easily solved. Metal ion oxidation by molecular oxygen induces the oxidation of hydrocarbons (alkanes among them) introduced in the system. The formation of alkane oxidation products, at least with low yields, has been observed in many cases even in the oxidation of simple ions in various media.

However, it is as yet premature to conclude that these reactions represent simple models of biological oxidation of hydrocarbons, if one considers the models as systems having a mechanism similar to that of biological oxidation.

The oxidation of organic compounds by dioxygen, including hydrocarbons, is known to be a strongly exothermic process with great negative free energy.

For a certain mechanism to correspond to an appropriate reaction rate, the intermediate stages must be advantageous enough thermodynamically, i.e. they should not require too much energy. The probability of this will be generally higher in the case of overall strongly exothermic processes, including oxidation of hydrocarbons by molecular oxygen, than in the case of less exothermic or thermoneutral processes. There are a number of mechanisms to provide a high rate of the whole process in the case of oxidation of hydrocarbons, particularly in coupled oxidation. They are mechanisms with chemically active species, such as free radicals, readily reacting with hydrocarbons. These free radicals may be formed as intermediates in metal ion oxidation. The mechanism with the participation of free radicals, at any rate such as OH or HO_2, is obviously far from being always acceptable for biological systems requiring substrate specificity. Moreover, these radicals would destroy the enzyme attacking different C—H bonds in its molecule.

The unique mechanism of monooxygenases considered above was evidently developed in nature to ensure the oxidation of only those compounds which are to be oxidized, even those as inert as methane and its analogues, not touching the versatile C—H bonds of the enzyme itself.

To realize this substrate specificity in the process of evolution, an enzyme system had to be created to expend 'unproductively' the energy of the donor in the coupled oxidation. Naturally, the energy of alkane oxidation is sufficiently high in itself and there is no need of the supplementary donor energy for the process to continue. The donor is spent to create a necessary catalytically active centre.

Therefore, one should check up accurately whether the mechanism of a model chemical reaction corresponds to the mechanism of an enzyme process.

The coupled oxidation was described first in the middle of the past century.

Schönbein discovered the phenomenon of a so-called 'active oxygen' [88, 89], i.e. oxidation by molecular oxygen of some substances only when there is parallel oxidation of other substances present. For example, the formation of iodine from potassium iodide or the discolouring of indigo are induced by zinc or zinc amalgam. Ferrous oxide is among the number of organic, and inorganic substances, whose oxidation induces the oxidation of more inert substances.

A thorough review of the oxidation reactions known at that time is given in the book by N. A. Shilov, which appeared in 1905 [90]. Several theories of active oxygen already existed then. According to N. A. Shilov, "All of them come to an assumption that the coupled oxidation by molecular oxygen is conditioned by the formation of *an intermediate product*. Some authors consider this intermediate substance to be free oxygen atoms or ions; the others, on the contrary, suppose that the oxygen molecule in the first stage of oxidation preserves the atomic complex —O—O—, forming this or that holoxide.* This last group of theories elucidates the experimental evidence in the best way".

The suggestion of peroxide formation in the first stage of the oxidation by molecular oxygen was made by Schönbein, their intermediate formation being confirmed in the works of Traube, Bach, Engler and others.

For the coupled oxidation of a number of substances by strong oxidizing agents (for example, H_2CrO_4 or $HMnO_4$) in the presence of ferrous oxide, as long as 80 years ago Manchot [91, 92] suggested an intermediate formation of high-valent iron oxide, Fe_2O_5. Somewhat analogous alternative hypotheses on the coupled oxidation (especially, for biological systems) are being discussed now, 80 years later.

According to Bach and Engler's peroxide theory [30, 31], the coupling in oxidation reactions is explained in terms of intermediate formation of peroxide. If of two substances, A and B, only one (e.g., A) reacts with molecular oxygen to give peroxide (AO_2) (substance A is then called 'inductor'), then substance B ('acceptor') may be oxidized by this peroxide:

$$B + AO_2 \longrightarrow AO + BO$$

However, later it became clear that in most cases the assumption of the stable peroxide cannot by itself explain the phenomena observed. For example, though benzaldehyde oxidation initiates the coupled oxidation of indigo, specially synthesized perbenzoic acid can hardly oxidize indigo, and if so, the reaction proceeds far slower than it does in the process of coupled oxidation.

_____

* A substance of the $R{<}{\begin{smallmatrix} O \\ | \\ O \end{smallmatrix}}$ type [Author].

The development of the chain theory has produced a simple explanation of these facts in terms of intermediate formation of radicals. The latter react with molecules much more readily than stable hydroperoxide molecules. The formation of free radicals was further established for many cases of oxidation of metal ions and complexes.

IV.4.1. HYDROXYLATION OF HYDROCARBONS COUPLED WITH OXIDATION OF METAL IONS AND COMPLEXES

The fact of metal ion participation in most enzyme systems, which hydroxylate organic compounds, was the reason behind many attempts to create analogous purely chemical systems.

In 1954 Udenfriend suggested a non-enzymatic system as a possible model of monooxygenase [93]. It involves an iron(II) complex with EDTA and ascorbic acid and is capable of hydroxylating aromatic compounds. Later Hamilton found that this system was able to hydroxylate cyclohexane (however, with very low yields) and epoxidize cyclohexene [73]. Afterwards, a number of similar systems were proposed.

For example, Ullrich found two models which are more effective than the Udenfriend system. One of them includes a Sn(II) phosphate complex dissolved in water [94], and the other one an iron(II) complex with mercaptobenzoic acid in aqueous solution of acetone [66].

Table IV.6. lists the products of some typical systems capable of coupled hydroxylation of hydrocarbons induced by metal ion oxidation by molecular oxygen. The ratio of isomers produced $(o:m:p)$ for the hydroxylation of aromatic compound derivatives is given in brackets.

TABLE IV.6

Systems involving ions or metal complexes and molecular oxygen [95]

System	Substrate	Products $(o:m:p)$
Fe^{2+} + ascorbate	Androsten-3-ol-17-one	Androsten-3,7-diol-17-one
Fe^{2+} + ascorbate	Cortexolone	Cortisol + cortisone
Fe^{2+} + ascorbate + EDTA	Anisole	Methoxyphenols 43:18:39
Fe^{2+} + 2,4,5-triamino-6-oxypyrimydine	Anisole	Methoxyphenols 49:13:38
Fe^{2+} + thiosalicylic acid	Acetanilide	Phenols 57:3:40
The same	Toluene	Cresols 61:13:16
The same	Naphthalene	α-Naphthol 11%
		β-Naphthol 9%

Table IV.6 (Continued)

System	Substrate	Products ($o : m : p$)
Fe^{2+} + N-benzene-1,4-dihydronicotinamide	ArH	ArOH
Fe^{2+} + tetrahydropterin	Phenylalanine	Tyrosine 2 : 1 : 1
The same	Tryptophane	5-Oxytryptophane + melanin + kynurenin
Hemin + thiosalicylic acid	Cyclohexane	Cyclohexanol
Fe^{2+} + EDTA	Toluene	Cresols 38 : 24 : 38
The same	Anisole	Methoxyphenols 56 : 8 : 36
The same	Fluorobenzene	Fluorophenols 2 : 45 : 53
$FeCl_2$ (acetone and methanol)	Cyclohexane	Cyclohexanol
$TiCl_3$	Acetanilide	Phenols 35 : 32 : 33
$TiCl_3$	Toluene	Cresols 61 : 16 : 23
$TiCl_3$	Anisole	Methoxyphenols 54 : 0 : 46
$TiCl_3$	Fluorobenzene	Fluorophenols 2 : 33 : 65
$TiCl_3$	Nitrobenzene	Nitrophenols 4 : 36 : 60
$TiCl_3$	Cyclohexane	Cyclohexanol + cyclohexanone
CuCl	Toluene	Cresols 29 : 29 : 49
CuCl	Anisole	Methoxyphenols 75 : 8 : 17
CuCl	Fluorobenzene	Fluorophenols 30 : 68 : 2
$SnHPO_4$	Toluene	Cresols 46 : 38 : 16
$SnHPO_4$	Anisole	Methoxyphenols 53 : 20 : 17
$SnHPO_4$ − organic solvent	Acetanilide	Phenols 34 : 44 : 22
$SnCl_2$ − acetonitrile	Alkanes	Isomeric alcohols
$SnCl_2$ − acetone	Cyclohexane	Cyclohexanol
$MoCl_3$	Cyclohexane	Cyclohexanol
$Mo(CO)_6$ − acetonitrile	Naphthalene	α-Naphthol
The same	Benzene	Phenol
$MoO(S_2CNR_2)_2$	PPh_3	$Ph_3P{-}O$

The reactions are analogous to those catalyzed by monooxygenase: hydroxylation of alkanes and aromatic compounds, and in some cases, epoxidation of olefins.

There are some other analogues of biological oxidation. Ascorbic acid in the Udenfriend system may be replaced by pyrimidine derivatives which resemble tetrahydropteridines, which may be used as donors both in the model and enzyme systems. The reaction occurs at room temperature at pH 7.

As mentioned above, Hamilton suggested that the model systems (at least some of them) interact with the substrates via a so-called oxenoid mechanism similar to that of monooxygenase functioning.

In the case of ascorbic acid this mechanism is represented by the following [73]:

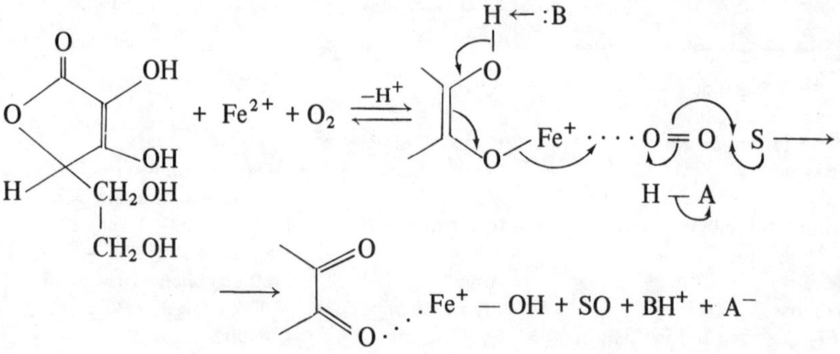

At the same time, there are some essential differences which may be noted between biological oxidation and processes occurring in the model systems. The latter have no NIH-shift (migration of hydrogen atom to the position next to the point of oxygen insertion), which is characteristic of biological systems in the hydroxylation of aromatic molecules. The model systems are mostly less selective in the reactions with aromatic compounds (a large amount of *meta*-substituted product is formed); they do not show stereospecificity in the hydroxylation of alkanes (e.g., a mixture of both possible isomers in the hydroxylation of *cis*- and *trans*-dimethylcyclohexane is formed).

Evidently, the participation of reducing agents, which are analogues of biological electron donors, is not very essential. In effect, simple chlorides of some metallic ions are already active enough in organic solvents.

Thus, the coupled oxidation of $C_1 - C_4$ alkanes and $SnCl_2$ was found to proceed in acetonitrile at room temperature [96]. The yields of alcohols, which are the oxidation products, reach 7–15% per $SnCl_2$.

Comparatively high yields of the products of cyclohexane oxidation are obtained in the presence of oxidizing tin(II) or iron(II) chlorides, when the reaction is carried out in acetonitrile as solvent [97] (Table IV.7).

The yield, if based on the metal salt present, may be further increased when the MCl_2 concentration is decreased. Under optimum conditions at $[MCl_2] \longrightarrow 0$, the yield of products grows up to 20% with $SnCl_2$ and up to 30% with $FeCl_2$.

Table IV.8 gives the distribution of isopentane hydroxylation products in various systems, including photochemical hydroxylation by hydrogen peroxide, where free hydroxyl radicals are known to be formed as intermediates. It is clear that the ratios of selectivities with respect to the site of attack ($1° : 2° : 3°$) in

TABLE IV.7

MCl$_2$	medium	[MCl$_2$] $M \times 10^2$	[C$_6$H$_{12}$] $M \times 10^2$	ΔO$_2$ $M \times 10^2$	[C$_6$H$_{10}$O] $M \times 10^3$	[C$_6$H$_{11}$OH] $M \times 10^3$	Yield on MCl$_2$ mol.%	Yield on C$_6$H$_{12}$ mol.%
SnCl$_2$	4ml CH$_3$CN 0.7ml H$_2$O 0.2 N HCl	10.0	19.0	5.5		16.0	16	8.5
FeCl$_2$	4ml CH$_3$CN 0.25 N HCl	4.0	23.0	2.1	0.6	3.25	9.6	1.7

the systems involving Fe^{2+} and Sn^{2+} ions are very close to each other and to the selectivity observed for the attack by hydroxyl. This leads to the conclusion [98] that the same active species (presumably hydroxyl radicals) participate in all these systems. This is also supported by a low isotope effect ($k_H/k_D \approx 1.2-1.3$, when comparing reactions of C$_6$H$_{12}$ and C$_6$D$_{12}$).

TABLE IV.8

System	Medium	Selectivity		
		prim.	*sec.*	*tert.*
heme + MBA* + O$_2$	CH$_3$COCH$_3$ + H$_2$O	1	5.7	14
Fe^{2+} + MBA* + O$_2$	CH$_3$COCH$_3$ + H$_2$O	1	4.1	12.8
SnCl$_2$ + O$_2$	CH$_3$CN + H$_2$O	1	5.1	12.5
FeCl$_2$ + O$_2$	CH$_3$CN	1	4.3	13.0
H$_2$O$_2$/$h\nu$	CH$_3$CN	1	4.5	11.2
CF$_3$CO$_3$H	RH	1	60	720
pyridine-*N*-oxide	CH$_2$Cl$_2$	1	8	68

* MBA – mercaptobenzoic acid.

The yield of cyclohexane hydroxylation products depends on the nature of the solvent (Table IV.9). The high yields in acetonitrile as compared with the other solvents are to be expected, provided the hydroxyl radicals are the intermediate active species. The rate constant of the hydroxyl radical reaction with cyclohexane is 5×10^9 M^{-1}s^{-1}, with acetonitrile 2.1×10^6 M^{-1}s^{-1}, with acetone 8.1×10^7 M^{-1}s^{-1}, and with methanol 7×10^8 M^{-1}s^{-1} [99]. In the case of cyclohexane in acetonitrile, this means that the reaction of ·OH

TABLE IV.9

The yield of products of cyclohexane oxidation coupled with oxidation of $SnCl_2$ and $FeCl_2$ in various solvents

System \ Solvent	The yield of products ($\times 10^3$ M) in solvents				
	Aceto-nitrile	Acetone	Ethyl acetate	Methanol	Water
$FeCl_2$* + O_2	4.6	1.6	0.5	0.1	0.1
$SnCl_2$** + O_2	8.0	3.0	0.8	0.1	0.1

* [HCl] = 0.17 M; [H_2O] = 7.0 M; [C_6H_{12}] = 0.23 M; [$FeCl_2$] = 0.17 M.
** [C_6H_{12}] = 0.23 M; [$SnCl_2$] = 0.35 M; [H_2O] = 0.7 M (in CH_3CN).

with the solvent at [C_6H_{12}] > 10^{-2} M may be neglected. For the other organic solvents it is necessary to take into account the competitive reaction of ·OH with solvent, which reduces the yield of cyclohexane hydroxylation products.

In the case of iron salts the accepted mechanism includes the formation of hydroxyl radicals

$$Fe^{2+} + O_2 + H^+ \longrightarrow Fe^{3+} + HO_2\cdot$$
$$Fe^{2+} + HO_2\cdot + H^+ \longrightarrow Fe^{3+} + H_2O_2$$
$$Fe^{2+} + H_2O_2 \longrightarrow Fe^{3+} + \cdot OH + OH^-$$
$$Fe^{2+} + \cdot OH \longrightarrow Fe^{3+} + OH^-$$
$$2\,H^+ + 2\,OH^- \longrightarrow 2\,H_2O$$

The total reaction: $4\,Fe^{2+} + O_2 + 4\,H^+ \longrightarrow 4\,Fe^{3+} + 2\,H_2O$

For stannous chloride the scheme of unbranched chain reaction without the participation of OH radicals has been proposed [100].

$$Sn^{2+} + O_2 \xrightarrow{H^+} Sn^{3+} + HO_2\cdot$$
$$Sn^{2+} + HO_2\cdot \xrightarrow{H^+} Sn^{3+} + H_2O_2$$
$$Sn^{3+} + O_2 \xrightarrow{H^+} Sn^{4+} + HO_2\cdot$$
$$Sn^{2+} + H_2O_2 \longrightarrow Sn^{4+}O^{2-} + H_2O$$
$$2\,Sn^{3+} \longrightarrow Sn^{2+} + Sn^{4+}$$

However, a detailed study of this reaction showed that it proceeds with chain branching, and that it involves hydroxyl radicals (see below).

The yield of oxidation products per metal ion may be increased if the reaction is carried out in the presence of specially added simple inorganic reducing agents. For example, if the coupled oxidation of $SnCl_2$ and C_6H_{12} is carried out in the presence of metallic tin, then the yield of cyclohexanol can be increased six times. For $FeCl_2$-containing systems, metallic mercury may be used as a reducing agent. This also increases the yield of hydroxylation products.

In the systems involving, besides metal compounds, such donors of electrons as hydrazobenzene or o-phenylenediamine, the yield of hydroxylation products may exceed the amount of the metal compound taken [101] (see Table IV.10).

TABLE IV.10

Hydroxylation of cyclohexane by the system MX/hydrazobenzene/O_2, t = 20°C, [cyclohexane] = 4.82 M, [PhNHNHPh] = 10^{-3} M, reaction time 3 h. [MX] = 5 × 10^{-3} M; [Acid] = 1 × 10^{-1} M.

Solvent	MX	Acid	Cyclo-hexanol, ($M \times 10^3$)	Cyclo-hexanone, ($M \times 10^3$)
Acetone	$FeCl_2$	–	4.05	0.1
''	''	benzoic	22.4	0.4
''	''	propionic	21.8	0.5
CH_3CN	''	benzoic	13.4	0.2
CH_3CO_2Et	''	''	11.7	0.3
CH_3OH	''	''	6.5	–
$C_2H_4Cl_2$	''	''	5.02	0.9
Acetone	$FeCl_3$	''	20.0	0.5
''	$FeBr_2$	''	12.0	0.4
''	FeI_2	''	2.0	–
''	$Fe(ClO_4)_2 6H_2O$	''	2.0	–
''	$FeCl_2 4py$	''	24.0	0.7
''	$FeCl_2 + \alpha$-naphthol	''	21.5	0.6
''	$FeCl_2 + H_2O_2$ (10^{-1} M)	''	0.2	0.7
''	$FeCl_2(bipy)_2$	''	7.3	0.2
''	$MnCl_2$	''	19.0	0.2
''	$MnBr_2$	''	19.5	0.1
''	VCl_2	''	17.2	0.2
''	$CuCl_2$	''	9.8	0.4
''	$NiCl_2$	''	5.2	0.5

Cyclohexane produces cyclohexanol and cyclohexanone, whereas cyclohex-1-ene-3-ol is mainly produced in the case of cyclohexene (without epoxide formation). Hydroxylation of 2-methylbutane produces isomeric alcohols, the selectivity with respect to the site of attack following the order: $3° > 2° > 1°$.

In toluene hydroxylation, the formation of both benzyl alcohol and isomeric cresols is observed, o- and p-isomers prevailing appreciably.

An oxenoid mechanism of hydroxylation has been suggested [101] (Scheme IV.4.1.1).

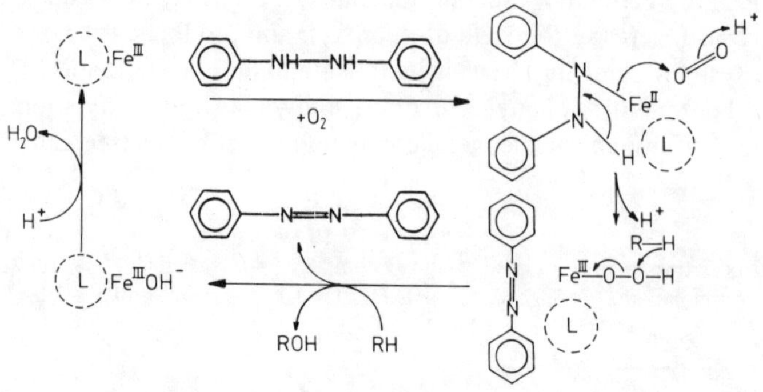

Scheme IV.4.1.1

However, it is difficult to reconcile this mechanism with the fact that the yield of cyclohexane hydroxylation products depends so insignificantly on the nature of the metal ion (for $FeCl_2$, VCl_2, $MnCl_2$, $NiCl_2$, $CuCl_2$ the yield differs no more than 4 times; while for $FeCl_2$, $MnCl_2$, VCl_2 it is almost the same). The formation of the same intermediate active species, for example, $HO_2 \cdot$, seems to be a more natural alternative.

A particularly active system was observed when mixing copper dichloride with phenylhydrazine in various organic solvents and water [26]. Under optimum conditions, the yield reaches up to 40% per oxidized phenylhydrazine, copper salt acting as a catalyst (Table IV.11). The yield of products may correspond,

TABLE IV.11

Cyclohexane oxidation in the $CuCl_2 + O_2$ + phenyl hydrazine system;
$t = 20°C$, solvent: 3.5 ml; cyclohexane: 1.5 ml; phenyl hydrazine:
0.1 M; $CuCl_2$: 1×10^{-3} M; v_0: initial rate.

Solvent	$v_0 \times 10^4$ M min^{-1}	Product yield (in % based on phenyl hydrazine)	
		Cyclohexanol	Cyclohexanone
Methanol	30	20	1.3
Ethanol	27	18	4
Acetone	1.6	32	10
Acetonitrile	10	42	6
Methyl chloride	2	12	0.6
Water	22	34	20

for example, to forty redox cycles per mole of Cu salt with one portion of the reducing agent, and even several hundred cycles, if new portions of phenyl hydrazine are added after the previous hydrazine has been oxidized. Evidently, this system — similarly to those previously reported — is not active with respect to methane and its close homologs.

Thus, a great number of species participating in the autoxidation of transition metal compounds makes it difficult to select a hydroxylation mechanism. Only in a small number of cases have this mechanism and the one of coupled hydrocarbon hydroxylation been well established. Let us take as an example the oxidation of stannous chloride.

IV.4.2. THE MECHANISM OF STANNOUS CHLORIDE AUTOXIDATION

Sn(II) is a typical two-electron reductant, since there are only two stable oxidation states for the compounds, that is, Sn(II) and Sn(IV). The redox potential of the pair Sn(IV)/Sn(II) is +0.15 V in aqueous solutions. Thus Sn(II) is a moderately strong reducing agent.

The estimation of redox potential of the pair Sn(III)/Sn(II) gives a value +0.8 V [102]. That means that E_0 for Sn(IV)/Sn(III) is −0.5 V, i.e. Sn(III) should be both a strong reducing agent turning into Sn(IV) and a strong oxidizing agent turning into Sn(II). The disproportionation of two Sn(III) ions is assumed to occur readily upon collision:

$$Sn(III) + Sn(III) \longrightarrow Sn(IV) + Sn(II) + \sim 30 \text{ kcal mole}^{-1}$$

The oxidation of tin(II) salts by dioxygen displays some of the features of a chain process; in particular, it is retarded by certain inhibitors and accelerated when the solution is irradiated.

The kinetic behaviour of the oxidation reaction of Sn(II) chloride has some peculiarities (Fig. IV.4). The reaction, under certain conditions, has a marked induction period followed by a period of practically constant rate, provided the dioxygen concentration is kept constant. The study of the dependence of the reaction rate on the dioxygen concentration shows that the reaction rate under steady-state conditions can be described by a simple kinetic expression

$$-\frac{d[O_2]}{dt} = k[O_2]^2$$

i.e. the reaction rate does not depend on the Sn(II) concentration. It is clear that such a kinetic expression, though very simple, cannot be explained by a simple

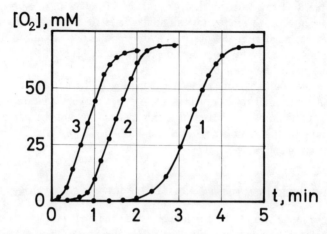

Fig. IV.4: The kinetic curves of autoxidation of the solution of $SnCl_2 \cdot 2H_2O$ (0.11 M) in tetrahydrofuran at 15°C and water content: (1) 0.45 M; (2) 1.1 M; (3) 2.2 M; $[O_2]$: amount of O_2 absorbed.

mechanism. To explain the kinetic data obtained, a branching chain scheme has been proposed [103, 104, 105]:

(0) $Sn(II) + O_2 \xrightarrow{H_2O} Sn(III) + HO_2 \cdot$ initiation.

(1) $Sn(III) + O_2 \xrightarrow{H_2O} HO_2 \cdot + Sn(IV)$

(2) $HO_2 \cdot + Sn(II) \Big\langle \begin{array}{l} \xrightarrow{\beta} H_2O_2 + Sn(III) \\ \xrightarrow{1-\beta} \cdot OH + Sn(IV) \end{array}$

(3) $H_2O_2 + Sn(II) \Big\langle \begin{array}{ll} \xrightarrow{\alpha} \cdot OH + Sn(III) & a \\ \xrightarrow{1-\alpha} H_2O + Sn(IV) & b \end{array}$

(4) $\cdot OH + Sn(II) \longrightarrow Sn(III)$

(5) $Sn(III) + Sn(III) \longrightarrow Sn(II) + Sn(IV)$

Here α and β are the probabilities of one-electron reactions of Sn(II) with H_2O_2 and $HO_2 \cdot$, respectively.

In steady-state conditions

$$\alpha\beta k_1 [Sn(III)] [O_2] = k_5 [Sn(III)]^2$$

and the reaction rate is practically equal to that of chain propagation

$$-\frac{d[O_2]}{dt} = k_1 [Sn(III)] [O_2] = \frac{\alpha\beta k_1^2}{k_5} [O_2]^2$$

and, in agreement with the experimental data, depends only on dioxygen concentration.

The branching chain mechanism is confirmed, for example, by the sensitivity of the reaction to the inhibitor concentration (Fig. IV.5). There exists some

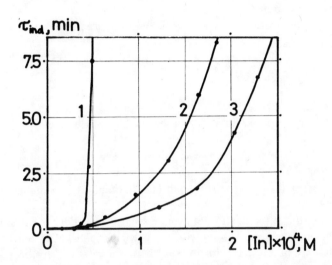

Fig. IV.5: The dependence of the induction period (τ_{ind}) in aquo-acetonitrile solutions on the inhibitors concentration: (1) naphthylamine ($[Sn(II)]_0 = 0.175$ M, $[H_2O] = 7.0$ M, air, $t = 20°C$); (2) tetranitromethane ($[Sn(II)]_0 = 0.11$ M, $[H_2O] = 6.7$ M, air, $t = 25°C$); (3) β-naphthoquinone ($[Sn(II)]_0 = 0.18$ M, $[H_2O] = 2.2$ M, air, $t = 20°C$).

limiting inhibitor concentration which practically stops the reaction (the induction period increases to infinity on the addition of a definite amount of an inhibitor). Hydrogen peroxide was detected during the oxidation of Sn(II) [103]. Using the reaction scheme, it was possible to estimate the rate constant of the reaction of hydrogen peroxide with Sn(II) by the dependence of H_2O_2 concentration on time. The rate constant found coincided with the value obtained for the reaction of H_2O_2 with Sn(II) determined in a separate investigation, thus confirming the proposed mechanism.

Though the probability (α) of a one-electron reaction of hydrogen peroxide with Sn(II) to give hydroxyl radical is not very high ($\alpha \approx 5 \times 10^{-4}$ for the reaction in acetonitrile), it is quite sufficient for the reaction to ensure the high rate of radical formation in a branched chain reaction under steady-state conditions.

The rate of autoxidation is much lower in acidic aqueous solutions of Sn(II) chloride than in organic solvents (though the addition of water is necessary because the OH and HO_2 radicals, as well as hydrogen peroxide, participate in the reaction, their formation requiring protons). The rate decrease in aqueous solutions is partially associated with the decrease of dioxygen solubility which is particularly important, since the O_2 concentration enters the kinetic equation as a second-order term.

However, the kinetic equation in acidic aqueous media is also different from that found for organic solvents [100].

$$v_{0_2} = k_{eff} [Sn(II)]^{1/2} [O_2]^{3/2}$$

Detailed studies [104, 105] have shown that the different form of the kinetic equation for aqueous solution is due to the reaction which takes place in water

(6) $Sn(III) + H_2O_2 \longrightarrow Sn(IV) + \cdot OH + OH^-$

and is nonessential for acetonitrile solutions. In other respects the mechanism is the same.

Therefore, the autooxidation of stannous salts in a variety of solvents occurs by a branching chain mechanism with the participation of $\cdot OH$ and $HO_2 \cdot$ radicals. Reactions of radicals with hydrocarbons should naturally be taken into consideration in the process of their oxidation coupled with the autoxidation of tin(II) salts.

IV.4.3. THE MECHANISM OF CYCLOHEXANE OXIDATION COUPLED WITH OXIDATION OF SnCl$_2$ [96–98, 104, 105]

On the introduction of cyclohexane into Sn(II) aqueous acetonitrile solution, the rate of dioxygen consumption increases and the induction period falls (Fig. IV.6).

An increase of cyclohexane concentration decreases the induction period and even causes its complete disappearance. The higher the initial tin concentration, the greater is the cyclohexane concentration necessary to cause the disappearance of the induction period. The rate of hydrogen peroxide production is also increased in the presence of cyclohexane. These data can be understood on the basis of the above chain mechanism of Sn(II) oxidation, where free hydroxyl

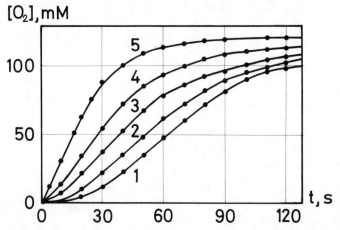

Fig. IV.6: The kinetics of the coupled oxidation of C_6H_{12} in aquo-acetonitrile solutions (t = 20°C, $[H_2O]$ = 1.4 M, $[Sn(II)]_0$ = 0.2 M, O_2 = 10% vol.) at the initial concentrations of C_6H_{12}: (1) 0.0 M, (2) 0.0047 M, (3) 0.023 M, (4) 0.09 M, (5) 0.47 M. $[O_2]$: amount of O_2 absorbed.

radicals $\cdot OH$ are intermediately formed. The rate constant of the interaction of hydroxyl radical with cyclohexane

$$\cdot OH + C_6H_{12} \longrightarrow H_2O + \cdot C_6H_{11}$$

is 5×10^9 $M^{-1}s^{-1}$, thus all the hydroxyl radicals, with an appropriate hydrocarbon concentration, will interact with the latter to produce hydrocarbon radicals. The radical very readily enters into the addition reaction in the presence of O_2.

$$R\cdot + O_2 \longrightarrow RO_2\cdot$$

In hydroperoxide oxidation reactions, the radicals usually disappear in a bimolecular disproportionation process, as mentioned earlier (p. 75):

$$2\,RO_2\cdot \longrightarrow R'COR'' + ROH + O_2$$

However, in the presence of large Sn(II) concentrations, practically all hydroperoxide radicals interact with the tin in a similar manner to HO_2 radicals

$$RO_2\cdot + Sn(II) \xrightarrow{\;H_2O\;} \begin{cases} \xrightarrow{\;\beta\;} RO_2H + Sn(III) \\ \xrightarrow{\;1-\beta\;} RO\cdot + Sn(IV) \end{cases}$$

The interaction of hydroperoxide with Sn(II) is similar to that of hydrogen peroxide

$$RO_2H + Sn(II) \quad \overset{\alpha}{\underset{1-\alpha}{\diagdown}} \quad \begin{array}{l} RO\cdot + Sn(III) \\ ROH + Sn(IV) \end{array}$$

RO· radicals react with RH and Sn(II)

$$RO\cdot + RH \longrightarrow ROH + R\cdot$$
$$RO\cdot + Sn(II) \xrightarrow{H_2O} ROH + Sn(III)$$

causing the formation of the alcohol, which is practically the only product. Besides, in aqueous solutions, as in the case of hydroperoxide, it is necessary to take into account the following reaction:

$$ROOH + Sn(III) \longrightarrow RO\cdot + Sn(IV)$$

Thus, the participation of Sn(II) in the oxidation process ensures high selectivity of the reaction to produce cyclohexanol as virtually the sole product. It is important that the situation changes, if the Sn(II) concentration is low; for example, if instead of dioxygen we add hydrogen peroxide to the Sn(II) solution. In this case, Sn(II) quickly disappears, and the reaction of $RO_2\cdot$ disproportionation becomes dominant, yielding a variety of products different from those observed for the oxidation by dioxygen. If we do not differentiate between the differences in reactions with high and low Sn(II) concentrations under the conditions of oxidation, it may lead us to an incorrect conclusion about the non-radical character of hydrocarbon oxidation coupled with Sn(II) oxidation by dioxygen.

It is interesting to note that, provided the RH concentration is sufficient for RO· radicals to disappear by the reaction with the hydrocarbon, the stoichiometry of the reaction under specified conditions corresponds to the ratio Sn(II) : O_2 : ROH = 1 : 1 : 1, which means that it will be the same as the stoichiometry of oxidation in the presence of monooxygenase.

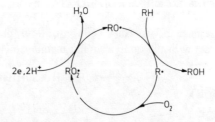

This shows that the stoichiometry of the process cannot be used as evidence for its mechanism. The example with Sn(II), treated here so extensively, is further evidence of the danger of drawing conclusions about the models of biological oxidation based merely on formal analogies.

IV.4.4. COMPLEXES OF MOLECULAR OXYGEN WITH TRANSITION METAL COMPOUNDS AND THEIR REACTIVITY

Complexes having O_2 in the metal coordination sphere are formed in the reaction of molecular oxygen with substances containing transition metals in a low oxidation state. These complexes are assumed to possess high reactivity and, in particular, to react with hydrocarbons. The complexes with O_2 are supposed to be intermediate active species in the biological oxidation of various substrates, for example, in the case of cytochrome P-450. As was mentioned above, one of the atoms of the metal-bonded oxygen molecule may be inserted according to the oxenoid mechanism in the C—H bond of the hydrocarbon or it may epoxidize olefins. Table IV.12 summarizes the examples of the well-known complexes of molecular oxygen [106].

Figure IV.7 depicts the structure of the octahedral complex with a 'picket-fence porphyrin' synthesized as a model complex for those formed in biological systems [107, 108]. Steric hindrances created by the presence of bulky groups in the porphyrin molecule prevent the destruction of the complex by the other iron complexes $Fe^{2+}O_2 + 3 Fe^{2+} + 2 H_2O \rightarrow 4 Fe^{3+} + 4 OH^-$.

A heme-mercaptido-O_2 complex, which is even closer to cytochrome P-450, has recently been synthesized, containing BuS^- in the coordination sphere of iron(II) [109a]. The complex is stable at $-50°C$ in aprotic solvents. Its electronic spectrum and other different characteristics are similar to those of cytochrome P-450, thus confirming the conclusion that $-S^-$ is the fifth ligand in the O_2 complex of cytochrome P-450 [109a, b]. Another similar O_2 complex stable at $0°C$ was reported recently [109c] with picket-fence perphyrin as a heme and tetrafluorophenyl mercaptide as an SH-containing ligand. An earlier model carbonyl ion porphyrin complex with mercaptide as a fifth ligand was synthesized [109d].

The diozygen complex is not active in the oxidation reactions of the C—H bond, and is not expected to be so, since one more electron has to be added to iron in the biological system in order to further activate the complex. Such a complex in solution has recently been obtained in a different way in a model reaction of a heme with a superoxide anion O_2^- [110a]. The complex has a structure of high-spin ferric porphyrin with η^2-peroxide and also exhibits no reactivity towards oxidation of alkanes, arenes and olefins [110b, c]. However, it can be activated in the presence of acylating agents (see Section IV.4.5).

TABLE IV.12

Compound	Type	Structure	O–O (Å)
O_2	free molecule	O–O	1.2074
KO_2	superoxide	O_2^-	1.28
H_2O_2	peroxide	O_2^{2-}	1.453
Co(bzacen)py·O_2 [a]	reversible O_2 complex	Co–O$\diagdown$O	1.26
'Picket fence' Fe(II) prophyrin methyl imidazole (O_2)	reversible O_2 complex	Fe–O$\diagdown$O	1.24
$[(NH_3)_5CoO_2Co(NH)_3]^{4+}$	peroxocomplex	Co–O$\diagdown$O–Co	1.47
$[(NH_3)_5CoOOCo(NH_3)_5]^{5+}$	oxidized peroxocomplex	Co–O$\diagdown$O–Co	1.31
$(O_2)IrCl(CO)[P(C_6H_5)]_3$	reversible peroxocomplex	O⸺O / Ir	1.30
$(O_2)IrI(CO)[P(C_6H_5)_3]_2$	irreversible peroxocomplex	O⸺O / Ir	1.51

[a] bzacen = N,N'-ethylene bis-(benzoylacetoniminide), $(C_6H_5-C(O^-)=CH-C(CH_3)=NCH_2-)_2$

The oxygen molecule in the coordination sphere of several other metal complexes displays some chemical reactivity. For example, SO_2, CO, NO, and N_2O_4 molecules are stoichiometrically oxidized by coordinated dioxygen to give the corresponding anions, which remain in the coordination sphere of the metal complexes, such as $IrX(CO)(O_2)P(C_6H_5)_3$ (where X = Cl, Br, I), $Pt(O_2)[P(C_6H_5)_3]_2$, $Ni(O_2)(tert\text{-}C_6H_9NC)_2$.

The catalytic oxidation of triphenylphosphine to triphenylphosphine oxide is also known in the presence of oxygen complexes with Ni(0), Pd(0), and Pt(0) compounds.

The superoxide anion O_2^-, coordinated in a complex with Ti(IV), reacts with different reductants (p-toluidine, p-aminophenol, aniline, phenol, etc.) but does not react with, for example, toluene, isopropyl alcohol, and benzaldehyde [111]. The authors of [111] have come to the conclusion that the oxidation by Ti(IV)O_2^- proceeds as a one-electron transfer from the donor to the acceptor rather than as an H atom abstraction since, while there is no correlation of the C–H bond dissociation energy, the reaction rate follows quantitatively the order of ionization potentials of the substrates. Thus, stable O_2 complexes known so far possess rather weak reactivity and in most cases are not able to interact with the alkanes.

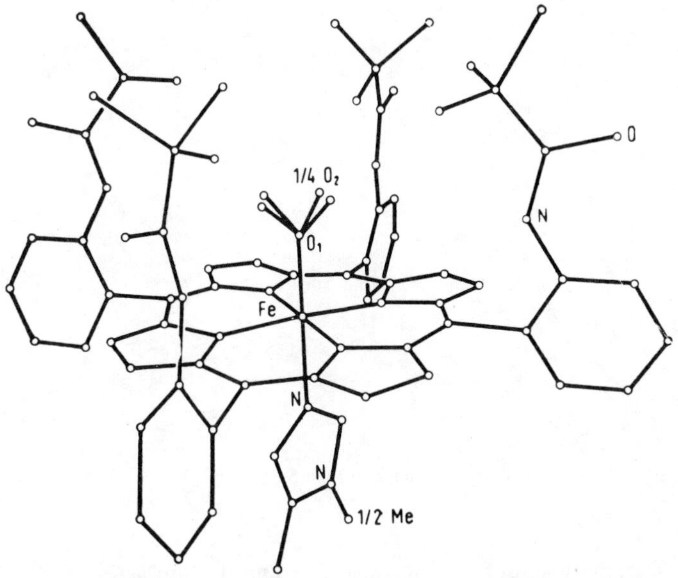

Fig. IV.7: The octahedral complex of Fe(II)O$_2$ with 'picket-fence porphyrin' as an equatorial ligand and methylimidazole as one of the axial ligands.

IV.4.5. PEROXIDES AND IODOSOBENZENE AS OXIDANTS: MODELS OF OXYGENASE ACTIVE CENTER

A number of systems for oxidizing hydrocarbons coupled with the oxidation of metal complexes are fully explained in terms of radical or radical-chain schemes. However, in many systems the mechanism is undoubtedly different. Presumably, non-aqueous and, particularly, aprotic media may be favourable for the hydroxylation mechanism involving active species based on a metal complex.

A study of cyclohexane cooxidation with ferrous chloride by dioxygen in ethanol allows us to conclude that, depending on the conditions, the hydroxylating species may be both OH and HO$_2$ radicals and such species as FeO_2^{2+}, as well as $FeOOH^{2+}$ and FeO^{2+} [112].

Hydroxylation by species other than OH and HO$_2$ free radicals is demonstrated in [113], where cyclohexanol is observed to be hydroxylated by hydrogen peroxide in the presence of ferrous perchlorate in acetonitrile. In this case, the product is mainly *cis*-1,3-cyclohexanediol (with more than 70% yield). Stereoselective oxidation was observed ot occur in this system for 7-hydroxynorbornane, too.

Scheme IV.4.5.1 is suggested for cyclohexanol hydroxylation, and it elucidates the formation of *cis*-1,4-cyclohexanediol.

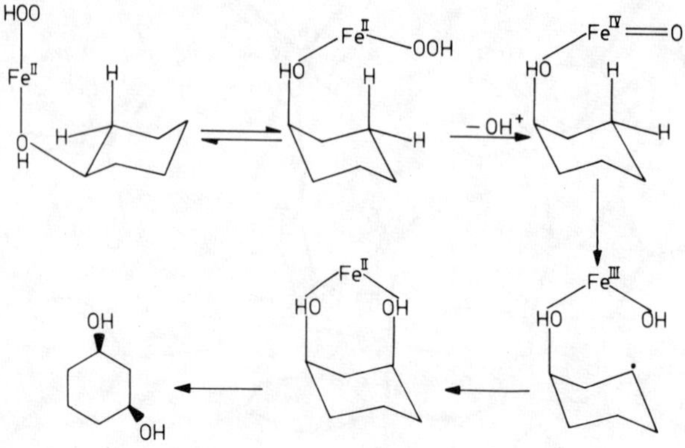

Scheme IV.4.5.1

Stereoselectivity is observed in the oxidation of hydrocarbons coupled with the oxidation of tin(II) organic derivatives (e.g., diphenyltin) in aprotic media [114]. In this case, cyclohexane oxidation produces cyclohexanol ($\sim$13% yield of diphenyltin). For the isopentane oxidation the ratio of selectivities $1° : 2° : 3°$ $\approx 1 : 16 : 107$ differs greatly from that for hydroxyl radicals. The hydroxylation of cis- and trans-dimethylcyclohexane proceeds with a partial retention of configuration in the alcohol formed (ca. 20%).

In the absence of the hydrocarbon, for example in CCl_4 solution, dioxygen forms an adduct with diphenyltin, the former obviously having a peroxide nature and being capable of hydrocarbon hydroxylation.

Molybdenum(VI) peroxide derivatives are known to epoxidize olefins [115]. The industrial process of olefin epoxidation by hydroperoxide is based on this reaction. The mechanism of epoxidation is believed to be the transfer of an oxygen atom to the double bond:

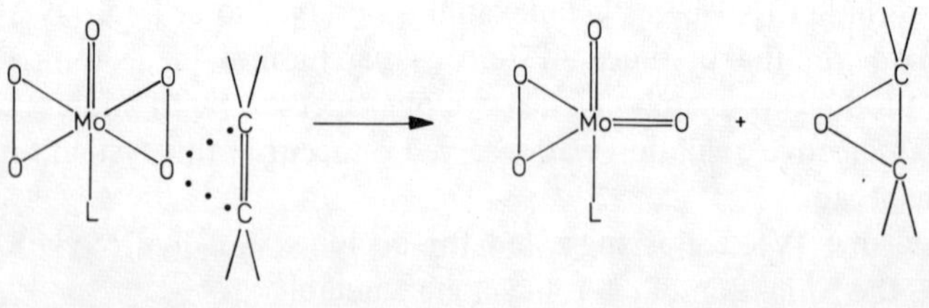

The molybdenum peroxocomplex acts as an electrophilic species with respect to the olefin.

Besides olefin epoxidation, molybdenum peroxides can hydroxylate aromatic and aliphatic hydrocarbons [116]. Thus, hydroxylation of naphthalene by a $MoO(O_2)L$ complex at 70°C in dichloroethane initially produces α-naphthol, which is further oxidized by the peroxocomplex.

Hydroxylation of alkanes by the molybdenum peroxocomplexes described in [116] evidently proceeds via the non-radical mechanism. In particular, the hydroxylation of *cis*- and *trans*-dimethylcyclohexane is stereoselective (50% retention of configuration is observed for *tert*-C—H bond hydroxylation).

Vanadium-containing peroxides, which are known to react with olefins to form epoxides, were recently found also to react with alkanes, forming ketones [117]. Thus, *tert*-butyl hydroperoxide in the presence of tetra-*tert*-butoxyvanadium(IV) reacts with *n*-octane and *n*-nonane at 20–100°C forming isomeric ketones. Primary C—H bonds in methyl groups are not affected. Presumably, vanadium-containing peroxides, e.g., $[VO(OBu)_2OOBu]^+$, are active in the reaction with alkanes.

It should be mentioned that boron and aluminium peroxides are known to be active towards alkanes, both reacting in accordance with the molecular mechanism [118–120].

It was stated (p. 97) that there is much evidence of the participation of high-valent iron $[FeO]^{3+}$ in the biological oxidation of hydrocarbons. Therefore, one may consider the reactions which occur under the action of high-valent metal compounds, such as

as mimicing, in a way, the biological oxidation [121].

In effect, the so-called NIH-shift which has earlier been found for biological hydroxylation (see p. 94) was observed in naphthalene oxidation by chromyl chloride in CCl_4 proceeding via the intermediate formation of α-naphthol.

A further approach to the biological systems has more recently been achieved with porphyrin complexes of several metals (Fe(III), Cr(V), Mn(III)) and iodosobenzene as an oxygen source [122–126]. The latter can be used also in the case of cytochrome P-450 enzymatic hydroxylation and epoxidation. A complete retention of configuration has been observed in the epoxidation of *cis*- and *trans*-stilbene with iodosobenzene using chlorodimethylferriprotoporphyrin as a catalyst, indicating that a stereoselective transfer of 'oxene' occurs, similar to the enzymatic epoxidation [122]. The aliphatic C—H bond was found to be oxidized by iron porphyrin complexes and iodosobenzene to give alcohols. Thus, cyclohexane afforded cyclohexanol in 8% yield.

At the same time, the absence of stereoselectivity in epoxidation by means of chloroaquo(tetraphenylporphinato)manganese(III) with iodosobenzene shows, apparently, that a free radical mechanism takes place in this case [125].

The first observation of a high-valent iron porphyrin complex which could be a close model of an intermediate in cytochrome P-450-catalyzed hydroxylation was reported recently by Groves *et al.* [126]. The complex was produced by the oxidation of chloro-5,10,15,20-tetramesitylporphinatoiron(III) with *m*-chloroperoxybenzoic acid in methylene chloride—methanol. It was identified by the NMR, visible, Mössbauer and ESR spectra as an iron(IV)-porphyrin π-cation radical structure (Fig. IV.8). The complex 1 can epoxidize olefins by transferring its oxygen atom to a double bond and regenerating the initial iron(III) complex.

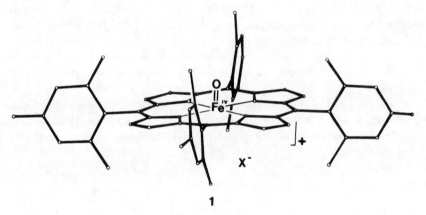

1

Fig. IV.8: A possible structure of the porphyrin cation-radical active in the oxidation of hydrocarbons.

The activation of oxygen in a peroxocomplex of ferric porphyrin may be achieved in model systems in the presence of acylating agents which presumably transform the peroxocomplex to the active form with one oxygen atom bound to Fe. Thus, during the addition of acetic anhydride to the solution of the peroxocomplex of ferric tetraphenylporphyrin obtained from the superoxide anion (see p. 119 in the previous subsection), cyclohexanol is produced in benzene—cyclohexane mixture [127].

The authors suggest that the formation of the active centre proceeds according to the scheme

$$[FeO_2]^+ + Ac_2O \xrightarrow[-OAc^-]{} [FeOOAc]^{2+} \xrightarrow[-OAc^-]{} [FeO]^{3+}$$

If this scheme is correct, the observed hydroxylation is a close model of oxidation by cytochrome P-450 (see p. 99).

As has been described earlier, there exists a variety of systems which can perform reactions similar to the reactions of monooxygenases without metal participation. They are, for example, the reactions of hydroxylation with the participation of peracids, in particular, trifluoroperacetic acid which is a strong electrophilic hydroxylating agent. It can epoxidize olefins, hydroxylate aromatic hydrocarbons and alkanes. The selectivity ratio of the C—H bond attack ($1°$: $2°$: $3°$ = 1 : 60 : 720) confirms the electrophilic properties of trifluoroacetic acid as a hydroxylating agent. There is a 100% retention of configuration in the reaction with *cis*- and *trans*-dimethylcyclohexane, an NIH-shift being observed in the hydroxylation of aromatic compounds.

Reactions of atomic oxygen also display a great resemblance to monooxygenases (stereoselectivity, NIH-shift).

Therefore, the biological oxidation of hydrocarbons with the participation of metal complexes is just a particular case of the widespread oxidation reactions with the C—H bond turning into C—OH.

IV.5. On The Mechanism of Alkane Reactions with Electrophilic Oxidants. General Considerations

It may be seen from the above examples that different mechanisms have been suggested in the literature for the reactions of saturated hydrocarbons with electrophilic oxidants. These various opinions reflect two factors. The first is the difficulty of solving the problem of the mechanism. No stable complexes of saturated hydrocarbons with metal compounds have yet been found, which can provide a certain key for the mechanism. The conclusions on the mechanism have to be based on primary products which are, in the case of alkanes, usually more reactive than the initial hydrocarbons. Hence, experimentally one usually observes some stable products, which are the result of a sequence of conversions, all this hampering the conclusion for the mechanism of initial hydrocarbon interaction.

The second factor, which makes it difficult to choose the mechanism of alkane reaction with metal compounds, is the real possibility of existence of a number of mechanisms depending on the nature of the reacting molecules, medium and reaction conditions.

The following probable mechanisms may be suggested, but we do not claim to be comprehensive.

1. *One-electron processes*

RH hydrocarbon reaction with a radical formed in the reaction of a transition metal compound, M, independently of the hydrocarbon

$$M \xrightarrow{\ x\ } R'\cdot$$
$$R'\cdot + RH \longrightarrow R'H + R\cdot \tag{IV.1}$$

One-electron transfer

$$M^{n+} + RH \longrightarrow M^{(n-1)+} + RH^+\cdot$$
$$\downarrow$$
$$H^+ + R\cdot \tag{IV.2}$$

One-electron transfer synchronous with proton elimination by the donor molecule B

$$M^{n+} + RH \xrightarrow{\ B\ } BH^+ + R\cdot + M^{(n-1)+} \tag{IV.3}$$

'Oxidative homolysis' with the participation of the ligand L

$$ML + HR \longrightarrow M + LH + R\cdot \tag{IV.4}$$

Reaction IV.3 with the addition of H^+ to the ligand L (or Reaction IV.4 with the elimination of an H atom with the help of the ligand) without LH leaving the coordination sphere of M

$$ML + RH \longrightarrow M(LH) + R\cdot \tag{IV.5}$$

2. *Two-electron processes*

Two-electron transfer

$$RH + M^{n+} \longrightarrow M^{(n-2)+} + [RH]^{2+}$$
$$\downarrow$$
$$R^+ + H^+ \tag{IV.6}$$

The elimination of a hydride-ion

$$M^{n+} + RH \longrightarrow [M-H]^{(n-1)+} + R^+ \tag{IV.7}$$

'Oxenoid' type reactions

$$M-OOH + RH \longrightarrow MOH + ROH$$
$$M-OO + RH \longrightarrow MO + ROH \tag{IV.8}$$
$$M=O + RH \longrightarrow M + ROH$$

(4) The electrophilic substitution followed by the reductive elimination

$$X^- M^{n+} + RH \longrightarrow X^-(M-R)^{(n-1)+} + H^+$$
$$\downarrow \qquad\qquad\qquad\qquad (IV.9)$$
$$M^{(n-2)+} + RX$$

Let us consider these mechanisms in the proposed sequence.

Radical formation in reactions of M independent of RH (IV.1.). As has been seen in many of the examples given in this section, this mechanism is widespread. Radicals may be formed in the unimolecular reaction of a metal compound (e.g., in the case of Mn(III) compounds) but more often in the reactions with the participation of intermediate species formed in the system. Here, various reactions with the participation of molecular oxygen are most important (Sections IV.2, IV.3), ions and complexes of transition metals serving as initiators of the hydrocarbon oxidation chain reaction or acting as catalysts (if cyclic changes of the metal oxidations state occur), or inducing the coupled hydrocarbon oxidation. The concentration of free radicals (such as OH, HO_2, RO_2, etc.) is maintained constant or even increases in the process of oxidation due to the energy provided by strongly exothermic reactions of the hydrocarbon or metal compound oxidation. This important general conclusion is easily traced with the example of the hydroperoxide participation in the chain reaction with degenerate branching. A hydroperoxide with the unstable O—O bond is formed intermediately, with the simultaneous formation of the stable C—O and O—H bonds. The further reactions of the hydroperoxide (with or without metal ions) create sufficiently high concentrations of free radicals, which makes the chain hydrocarbon conversion possible.

One-Electron Transfer (IV.2.). The possibility of the electron transfer from the hydrocarbon molecule to the transition metal compound molecule is not in any doubt whatsoever. However, the ionization potentials of saturated hydrocarbons are very high (see Table I of the Introduction), and it may be expected that very strong electron acceptors are required for the reaction to be possible at normal conditions. The redox potentials of some hydrocarbons may be calculated from their ionization potentials, if the solvation free energies of the respective ion-radicals are known. We can also use the calculated values for CH_3^+, $C_2H_5^+$, iso-$C_3H_7^+$, and tert-$C_4H_9^+$ ions [128] and consider solvation heats and entropies of these ions to be only slightly different from those of ion-radicals (CH_4^+, $C_2H_6^+$, etc.).

The radox potential of the $RH \rightleftharpoons RH^+ + e$ process can be calculated by comparing its free energy, which is equal to $I_{(RH)} + \Delta G_{(RH^+)}$, with that of the

$\frac{1}{2}H_2 \rightleftharpoons H^+ + e$ process (taken as zero) and equal to $I_H + \frac{1}{2}D(H-H) - \Delta G^S_{(H^+)}$, where I_{RH} and I_H are the ionization potentials of RH and H atoms, $\Delta G^S_{(RH^+)}$ and $\Delta G^S_{(H^+)}$ are the solvation free energies of RH^+ and H^+, and $D(H-H)$ is the H—H dissociation energy.

$$E = (RH/RH^+) = I_{RH} + \Delta G^S_{(RH^+)} - I_H - \frac{1}{2}D(H-H) + \Delta G^S_{(H^+)}$$

Table IV.13 gives the values obtained in this way. The calculated values for aromatic hydrocarbons are also given for comparison.

TABLE IV.13
Rexox potentials for the $RH \rightleftharpoons RH^+ + e$ processes in
aqueous solutions [129]

RH	$I(v)$	$-(\Delta G)_{solv.}$ kcal mole^{-1}	$E_0(V)$
CH_4	12.71	62	5.57
C_2H_6	11.5	54	4.68
C_3H_8	11.07	50	4.42
iso-C_4H_{10}	10.55	40	4.34
C_6H_6	9.25	56	2.45
$C_6H_5CH_3$	8.82	51	2.24

It is seen that for the alkanes, the redox potentials are apparently too high, and the reaction of simple electron transfer is evidently improbable even for the strongest oxidants, which would rather decompose water than react with the hydrocarbons. At the same time, this reaction is possible, in the case of sufficiently strong oxidants, for aromatic hydrocarbons, particularly for those involving donor groups and also for polycyclic aromatic hydrocarbons having a low ionization potential. Thus, in the absorption of perylene on the surface of the oxidation catalysts $MoO_3-Al_2O_3$ and MoO_3-SiO_2, perylene cation-radicals are formed which can be detected by ESR [130]. In this electron transfer process Mo^{6+} is reduced to Mo^{5+}

$$ArH + Mo^{6+} \longrightarrow ArH^+ + Mo^{5+}$$

Therefore, here we come across the reaction which proceeds readily in the case of some aromatic hydrocarbons, but is practically improbable for the alkanes.

One-Electron Transfer with Synchronous Proton Elimination (IV.3.). Proton elimination from RH^+ ion-radical is a thermodynamically favourable process, at least in aqueous solutions, since the difference between solvation heats of H^+ and RH^+ is considerably higher than the C—H bond energy in the alkanes. That is why the redox potentials for the $RH \rightleftharpoons R \cdot + H^+ + e$ reaction are much lower than for the $RH \rightleftharpoons RH^+ + e$ one.

The redox potentials for Reaction IV.3 for the hydrocarbon series can be calculated when comparing the two reactions:

$$\tfrac{1}{2}H_2 \rightleftharpoons H^+ + e \qquad E_1 = 0$$

$$RH \rightleftharpoons R \cdot + H^+ + e \qquad E_2$$

$$E_2 = \Delta H_{RH} - \Delta H_R.$$

Table IV.14 summarizes the values obtained in this way for the redox potentials of the hydrocarbon series. Evidently, these values are much lower than those for the formation of the RH^+ ion-radicals, and they are realistic even for the hydrocarbons of the methane series, provided sufficiently strong oxidants are used.

TABLE IV.14

Redox potentials of hydrocarbons for the $RH \rightleftharpoons R \cdot + H^+ + e$ processes in aqueous solutions ($pH = 0$). (The differences in $R \cdot$ and RH solvation energies are neglected [129])

RH	R	$E(V)$
CH_4	CH_3	2.25
C_2H_6	C_2H_5	1.99
C_3H_8	n-C_3H_7	1.99
C_3H_8	iso-C_3H_7	1.84
iso-C_4H_{10}	$tert$-C_4H_9	1.78
C_6H_6	C_6H_5	2.48
$C_6H_5CH_3$	$C_6H_5CH_2$	1.44
C_6H_{12}	C_6H_{11}	1.85

It should be noted that radical formation may even be an endothermic process and still provide a sufficient rate of radical formation, especially at elevated temperatures.

The presence of an oxidant having a redox potential higher than +1.5 V may result in the oxidation reaction in accordance with Mechanism IV.3. Since the electron transfer (from RH to M^{n+}) synchronous with the proton transfer (from RH to the base, e.g., water) must not be too complicated (taking into consideration the steric factors), this mechanism should be considered as very probable for various reactions of electrophiles-oxidants with hydrocarbons.

The fact of the co-participation of basic molecules in the reactions of proton elimination, which proceed synchronously with the other reactions, is quite common. The hydrocarbon reaction with the NO_2^+ cation is a close analogue of Reaction IV.3, where the reaction rate passes through a maximum with an increase of acidity [10, 28]. The rate decrease with high acid concentrations is thought to be associated with the decrease of the concentration of basic molecules which can participate in the reaction of proton elimination or electron transfer from the hydrocarbon molecule.

A sufficient redox potential of a transition metal compound can be reached with the addition of such strong oxidants as MnO_4^-, CrO_4^{2-}, Mn^{3+}, Co^{3+}, etc.

At first sight, the redox potential (E_0) of molecular oxygen is not sufficient, since for the four-electron conversion into water

$$O_2 + 4\,e + 4\,H^+ \rightleftharpoons 2\,H_2O$$

the $E_0 = +1.229$ V, for the one-electron process

$$O_2 + e + H^+ \rightleftharpoons HO_2\cdot$$

$E_0 = -0.13$ V, and for the two-electron conversion

$$O_2 + 2\,e + 2\,H^+ \rightleftharpoons H_2O_2$$

$E_0 = +0.682$ V.

However, the redox potential of species participating in the redox processes together with O_2 can be much higher than the redox potential of dioxygen itself.

The active species, which are chain carriers, can have high potentials, sufficient for the reaction with the hydrocarbons: for example hydroxyl radical, as has already been pointed out, can participate in the chain in the transition metal complex oxidation in aqueous solutions, and the redox potential for the reaction

$$OH\cdot + e \xrightleftharpoons{+H^+} H_2O$$

$E_0 = +2.8$ V.

As was mentioned above, the potential increase is here achieved due to the strongly exothermic nature of the process. There could be also a somewhat different mechanism of formation of intermediates with a high redox potential in reactions with the participation of dioxygen. O_2 is a four-electron oxidant; therefore, if there is a species capable of transferring four electrons to O_2, then a conjugated multielectron oxidant formed in the reaction may be, in principle, a very strong one- or two-electron oxidizing agent.

For example, in the monooxygenase participating reaction, if the process, as was suggested, proceeds according to the scheme

$$Fe^{3+} + O_2 + 2\,e \xrightarrow{(+2\,H^+)} FeO^{3+} + H_2O$$

then the $[FeO]^{3+}$ complex may be considered a product of a four-electron oxidation by dioxygen of Fe(I).

$$Fe^+ + O_2 + 2\,H^+ \rightleftharpoons FeO^{3+} + 2\,H_2O \qquad (IV.10)$$

Reaction IV.10 may be divided into three steps, as follows:

$$Fe^+ \rightleftharpoons Fe^{3+} + 2e \qquad 2\,E_1^0$$
$$Fe^{3+} + H_2O \rightleftharpoons FeO^{3+} + 2\,H^+ + 2\,e \qquad 2\,E_2^0 \qquad (IV.11)$$
$$O_2 + 4\,e + 4\,H^+ \rightleftharpoons 2\,H_2O \qquad 4\,E_{H_2O}^0$$

Thus ΔG of IV.10 is equal to

$$\Delta G_{IV.10} = 2E_1^0 + 2E_2^0 - 4E_{H_2O}^0$$

and

$$E_2 = \frac{1}{2}(\Delta G - 2E_1^0 + 4E_{H_2O}^0)$$

If both $\Delta G_{IV.10}$ and E_1 are not far from zero, which is likely to be the case, then E_2^0 (for the pair $[FeO]^{3+}/Fe^{3+}$) (Reaction IV.11) will be about twice as high as $E_{H_2O}^0$, i.e. $E_2^0 \approx +2.5$ V.

Such an oxidant will be strong enough to oxidize the alkanes in the reactions of type IV.3 of the one-electron process with a synchronous proton elimination.

'Oxidative Homolysis', ML + HR → M + LH + R (IV.4.). As has already been mentioned, this mechanism is suggested for the alkane reactions with the complexes of transition metals in acidic media; for example Pd(II) reactions in sulfuric acid solutions [10, 28].

Such a mechanism can be plausible in the case when the ligand is rather weakly bound to the metal, the value of the L—H bond energy being rather high. The reaction of hydrocarbons with molecular fluorine can be considered as an analogue of such a process

$$F_2 + HR \longrightarrow F\cdot + HF + R\cdot$$

This reaction is responsible for the initiation of chain fluorination of hydrocarbons. The reaction is practically thermoneutral for methane and slightly exothermic for other hydrocarbons. Reactions of such a type proceed with a low activation energy barrier [131]. However, reactions of molecular fluorine are rather an exception than a rule for H atom elimination by other molecules from saturated hydrocarbons. In effect, considering the C—H bond energies and a competition of the three processes

(a) $ML \longrightarrow M\cdot + L\cdot$

(b) $ML + HR \longrightarrow M + LH + R\cdot$

(c) $ML^{n+} + HR \longrightarrow ML^{(n-1)+} + R\cdot + H^{+}$

one can easily see that a comparatively rare correlation of energy of the $D(M—L)$ and $D(L—H)$ bonds should exist for Reaction b to become preferable. If $D(M—L)$ is too high, Reaction c will occur; if it is too small, a unimolecular Reaction 1 will be the fastest process.

It should be noted that the $D(L—H)$ bond energy must be at least close to the large $D(C—H)$ bond energy, which is often far from the case.

We could see, with a 'cobalt-bromide' catalysis as an example (Section IV.2.3), that the reaction prefers to proceed via a $Br\cdot$ atom formation and a further reaction with a hydrocarbon, rather than in a bimolecular reaction of Co(III) with RH. Evidently, the two processes

$$Co^{III}—Br \longrightarrow Co^{II} + Br\cdot$$

$$Br\cdot + RH \longrightarrow HBr + R\cdot$$

proceed without serious steric hindrances, whereas the bimolecular process of $Co^{III}—Br$ interaction with RH (HBr formation or electron transfer) must be, in any case, a sterically hindered reaction.

(IV.5). The situation may appear more preferable for this mechanism, which can be considered as a particular case of both Mechanisms IV.3 and IV.4. It is the reaction in which a proton (or a hydrogen atom) of a hydrocarbon molecule is added to the ligand L, which at the same time remains bonded to the metal atom M

$$ML + RH \longrightarrow MLH + R\cdot$$

This reaction may be regarded as an electron transfer from RH to ML with a simultaneous proton transfer to the ligand L, which here plays the role of a donor (a particular case of Mechanism IV.3), or as an atom H transfer to the radical-like ligand L (a particular case of Mechanism IV.4). The reaction of $L_{n}MO$ and $L_{n}MOO$ with RH are examples of such a reaction:

$$L_{n}MO + HR \longrightarrow L_{n}MOH + R\cdot \qquad (IV.12)$$

$$L_{n}MOO + HR \longrightarrow L_{n}MOOH + R\cdot \qquad (IV.13)$$

(where the ligands O and O_2 are considered to be O^{2-} and O_2^{2-} donors in the metal coordination sphere or radical-type particles M—O· and M—O—O·). The O—H bond can be more stable than C—H, which makes this reaction thermodynamically possible. The hydrogen atom elimination from the hydrocarbon molecule does not result in the M—O bond cleavage, i.e., in this case, a unimolecular splitting of the M—O bond would not compete with a bimolecular reaction with a hydrocarbon molecule.

It is interesting, from the point of view of Reactions IV.12 and IV.13, to compare reactions of alkanes with simple radicals containing one or two oxygen atoms, i.e. HO· and HO_2·.

L_nMO· may be regarded as a derivative of the HO radical, whereas L_nMOO· as an analogue of the HO_2 radical. It is seen from the data of Table IV.15 that the reaction of HO· with CH_4

$$HO· + CH_4 \longrightarrow H_2O + CH_3$$

is thermodynamically favourable both in the gas phase and in water solution, while for HO_2 radicals a similar reaction is strongly endothermic and thus virtually impossible at low temperatures. At the same time, the reaction of HO_2· with a tertiary C—H bond is almost thermoneutral and therefore quite probable.

TABLE IV.15

R·	$D(RH)$ kcal mole^{-1}	$E_{RH/R·+H^+}$ V
HO·	118	+2.8
HOO·	88	+1.5
CH_3·	104	+2.25
iso-C_4H_{10}	89.5	+1.78

We may expect that, though bond energies and redox potentials will be naturally influenced by iron and its surrounding, it is easier to construct active centres capable of attacking primary C—H bonds based on L_nMO species rather than on peroxo L_nMO_2 species. The latter, however, particularly those activated by electrophilic coaction of high-valent metal ion, can still be active in the case of more polar and less strong tertiary C—H bonds.

Two-Electron Reactions

Two-electron oxidation is usually more favourable thermodynamically than one-electron processes. The redox potentials for a particular case of two-electron alcohol formation are given in Table IV.16.

TABLE IV.16

The redox potentials of hydrocarbons for the
$RH + H_2O \rightleftharpoons ROH + 2\,e + 2\,H^+$ process

RH	ROH	$E(v)$
CH_4	CH_3OH	0.59
C_2H_6	C_2H_5OH	0.48
C_3H_8	$n-C_3H_7OH$	0.44
C_3H_8	$i-C_3H_7OH$	0.37
$iso-C_4H_{10}$	$t-C_4H_9OH$	0.31
C_6H_{12}	$C_6H_{11}OH$	0.30
C_6H_6	C_6H_5OH	0.20

These values are seen to be much smaller than those presented above for all the cases of one-electron oxidation.

Thus, if such a two-electron mechanism is possible, then in order to achieve it, significantly weaker oxidants can be used, whereas with oxidants of equal strength, it would correspond to a larger energy gain which must, in principle, ensure a higher rate. However, two-electron mechanisms may be kinetically difficult for various reasons.

A synchronous two-electron transfer (IV.6.):

$$M^{n+} + RH \longrightarrow R^+ + H^+ + M^{(n-2)+}$$

is evidently of low probability, since it requires a synchronous solvation of two positive ions by solvent molecules. The mechanism with a simultaneous cleavage of a C—H and a double M=O bond to form —O—R and O—H bonds, for example,

$$M \overset{..\,O}{\underset{..\,O}{<}} + RH \longrightarrow M \overset{OR}{\underset{OH}{<}} \longrightarrow M{=}O + ROH$$

with a decrease of the metal atom's oxidation state by two units also does not seem very probable, since it evidently requires a great distortion of the O—O interatomic distance. It might also be prohibited by the molecular orbital symmetry conservation rule.

Hydride-ion abstraction (IV.7.):

$$M^{n+} + RH \longrightarrow [M{-}H]^{(n-1)+} + R^+$$

is undoubtedly possible, since such a reaction is known for the case of the inter-action of alkanes with positive carbenium ions:

$$R'^+ + RH \longrightarrow R'H + R^+$$

However, the bond with H^- in the coordination sphere of the metal is probably in most cases much weaker than with R^+, so that such a reaction would be usually thermodynamically forbidden.

'Oxenoid' reactions (IV.8.):

(a) $MOOH + RH \longrightarrow MOH + ROH$

(b) $MOO + RH \longrightarrow MO + ROH$

(c) $MO + RH \longrightarrow M + ROH$

are thermodynamically favourable for types (a) and (b) and also favourable for type (c), at least when M is in a high oxidation state.

   These reactions may in fact encompass different mechanisms. Reactions of type (a), similarly to reactions of peracids (such as $CF_3 C(O)OOH$), may proceed by electrophilic action of metal hydroperoxide according to the molecular mechanism.

   Types (b) and (c) involve a radical reaction of H atom abstraction with a subsequent fast reaction of the radicals formed in a 'cage' of solvent. In this case both the (a) and (b) types will be more likely with tertiary and secondary C—H bonds while a primary C—H bond will be non-reactive.

   Presumably reactions of the 'true' oxenoid mechanism (similarly to CH_2 insertion) also start as H atom abstraction. Thus $L_n MO$ species must be again more active than $L_n MOO$. Therefore, in all the cases where primary C—H bond is inserted, it is $L_n MO$ rather than $L_n MO_2$ which probably constitutes the active center.

   Type (c) reactions are particularly interesting, since they may take place in biological oxidation of CH_3 groups in alkanes. For metal compounds in high oxidation state (e.g., FeO^{3+}) it is still difficult to distinguish this process from Reaction IV.12 of H atom abstraction followed by an immediate reaction of radicals formed in a 'cage'. Possibly both the mechanisms, i.e., the intermediate radical formation and the synchronous O atom insertion, may take place under different conditions and only close examination will allow a definite conclusion to be drawn.

Electrophilic substitution with the further two-electron oxidation (IV.9.):

$$X^- M^{n+} + RH \longrightarrow X^- (M-R)^{n-1} + H^+$$
$$\downarrow$$
$$M^{(n-2)+} + RX$$

must evidently be a thermodynamically favourable reaction in the case of oxidants capable of accepting two electrons. With respect to Reaction IV.3 (or Reactions IV.5, IV.11 and IV.12) of R radical formation this reaction, in its first stage, is more exothermic (or less endothermic) by a value of $D(M—R)$ bond energy, the subsequent reaction requiring a much weaker oxidant.

Steric effects must be a natural hindrance for this two-electron process. The appearance of steric hindrances indicated, for example, by abnormal selectivity at the attack of primary, secondary and tertiary C—H bonds may be an indirect evidence for the two-electron non-radical mechanism. However, as we could see for the reactions of alkanes with aminium radicals, the steric hindrances can arise with the radical mechanism as well.

Along with the unusual selectivity at the site of attack, an unusual stereo-selectivity is sometimes observed (particularly in biological oxidation), which might also lead us to the assumption of the non-radical and, particularly, the two-electron mechanism. As we shall see in the next section, the alkanes are capable of reacting with metal complexes to yield alkyl derivatives of the metal. Hence, Mechanism IV.9 of electrophilic substitution can take place in the stereo-selective oxidation and the oxidation with abnormal selectivity.

The foregoing equilbrium addition of alkanes, followed by proton abstraction, is expected for the case of strong acids. The cases with high isotope effect ($k_H/k_D = 4 - 5$) observed for some processes of non-biological oxidation (see Table IV.2) and biological oxidation with the participation of monooxygenases (e.g., methanemonooxygenase) might be examples of such a mechanism. The results showing small isotope effects may be interpreted, then, in terms of syn-chronous mechanisms (simultaneous electron and proton transfer or heterolytic electrophilic substitution with a synchronous formation of the M—C and a splitting of the C—H bond (of Type IV.3)). However, these conclusions are so far rather indirect. At present, the problem of existence of mechanisms with two-electron oxidation (for the reactions of alkanes with electrophiles-oxidants), in particular, the intermediate formation of metal alkyl derivatives for such reactions remains unsolved, and in no case have we direct evidence for such a mechanism.

The possibility of such a process follows mainly from the results of the study of the reactions of alkanes with platinum metal complexes (Chapter V), where the formation of the metal-carbon bond is well established.

As will be seen, the selectivity of the Pt(II) complexes' attack on various C—H bonds: $1° > 2° \gg 3°$ is analogous to the selectivity of the hydroxylation of alkanes from methanemonooxygenase. However, it is always dangerous to extrapolate by analogy. The reactions of alkanes with electrophilic oxidants differ drastically in their characteristics from the reactions with complexes of platinum and other metals of medium and low oxidation state. The final chapter will be devoted to reviewing these latter reactions.

References

1. R. Stewart: *Oxidation Mechanisms*, W. A. Benjamin, Inc., New York, Amsterdam (1964).
2. S. Sundaram, N. Venkatasubramanian, and S. V. Anantakrishnan: *J. Sci. Ind. Res.*, 35, 518 (1976).
3. A. Étard: *Ann. chim. et phys.*, 22, 218 (1881).
4. J. Roček and F. Mares: *Coll. Czechoslovakian Chem. Commun.*, 24, 2741 (1958).
5. K. B. Wiberg and G. Foster: *J. Amer. Chem. Soc.*, 83, 423 (1961).
6. J. Roček: *Tetrahedron Lett.*, 135 (1962).
7. R. Stewart and U.A. Spitzer: *Canad. J. Chem.*, 56, 1273 (1978).
8. V. P. Tretyakov, L. N. Arsamaskova, and Yu. I. Yermakov: *Kinet. katal.*, 15, 538 (1974).
9. L. N. Arsamaskova, Yu. I. Yermakov, V. A. Likholobov, and V. P. Tretyakov: *React. Kinet. Catal. Lett.*, 3, 183 (1975).
10. E. S. Rudakov and A. I. Lutsyk: *Neftekhimia*, 20, 163 (1980).
11. L. N. Arasmaskova, A. V. Romanemko, and Yu. I. Yermakov: *React. Kinet. Catal. Lett.*, 13, 391 (1980); idem: ibid, 13, 395 (1980).
12. E. S. Rudakov, V. P. Tretyakov, L. A. Minko, and N. A. Tishchenko: *React. Kinet. Catal. Lett.*, 16, 77 (1981).
13. R. A. Sheldon and J. K. Kochi: *Oxidation and Combustion Revs.*, 5, 135 (1973); R. A. Sheldon and J. K. Kochi: *Metal-Catalyzed Oxidations of Organic Compounds*, Acad. Press, New York-London-Toronto-Sydney-San Francisco (1981).
14. D. Benson: *Mechanisms of Oxidation by Metal Ions*, Elsevier, Amsterdam (1976).
15. J. K. Kochi: *Organometallic Mechanisms and Catalysis*, Academic Press, New York (1978).
16. T. A. Cooper and W. A. Waters: *J. Chem. Soc. (B)*, 687 (1967).
17. R. Tang and J. K. Kochi: *J. Inorg. Nucl. Chem.*, 35, 3845 (1973).
18. J. Hanotier, Ph. Camerman, M. Hanotier-Bridoux, and P. de Radzitzsky: *J. Chem. Soc., Perkin II*, 2247 (1972).
19. J. J. Ziolkowski, F. Pruchnik, and T. Szymanska-Buzar: *Inorg. Chim. Acta* 7(3) 473 (1973).
20. J. J. Ziolkowski and W. K. Rybak: *Bull. Acad. Pol. Sci. Ser. sci. chem.*, 22, 895 (1974).
21. L. Verstraelen, M. Lalmand, A. J. Hubert, and Ph. Teyssié: *J. Chem. Soc., Perkin II*, 1285 (1976).
22. S. R. Jones and J. M. Mellor: *J. Chem. Soc., Perkin II*, 511 (1977).
23. A. Onopchenko and J. G. D. Schultz: *J. Org. Chem.*, 40, 3338 (1975).
24. R. E. Partch: *J. Amer. Chem. Soc.*, 89, 3662 (1967).
25. S. R. Jones and J. M. Mellor: *J. Chem. Soc., Perkin I*, 2576 (1976).
26. V. V. Lavrushko, A. M. Khenkin, and A. E. Shilov: *Kinet. katal.*, 21, 276 (1980).
27. J. V. Crivello: *Amer. Chem. Soc., Div. Org. Chem. Meet. Chicago Pap*, No. 142 (1970).
28. E. S. Rudakov: *Izv. Sib. Otd. Akad. Nauk SSSR, ser. khim.*, 161 (1980).
29. N. F. Goldshleger, M. L. Khidekel, A. E. Shilov, and A. A. Shteinman: *Kinet. katal.*, 15, 261 (1974).
30. A. Bach: *Zh. Ross. Fiz. Khim. Obshch.*, 29, 25 (1897).
31. C. Engler and W. Wild: *Ber.* 12, 1669 (1897).
32. N. N. Semenov: *Chemical Kinetics and Chain Reactions*, Clarendon Press, Oxford (1935).

33. N. M. Emanuel, E. T. Denisov, and Z. K. Maizus: *Liquid-Phase Oxidation of Hydrocarbons*, Plenum Press, New York (1967).
34. C. Walling: *J. Amer. Chem. Soc.*, **91**, 7590 (1969).
35. F. Haber and R. Willstätter: *Ber.*, **64B**, 2844 (1931).
36. Z. K. Maizus, I. P. Skibida, and A. B. Gagarina: *Zh. Fiz. Khim.*, **49**, 2491 (1975).
37. C. E. H. Bawn and J. Jolley: *Proc. Roy. Soc.*, **237A**, 297 (1956).
38. C. E. H. Bawn: *Disc. Faraday Soc.*, **14**, 181 (1953).
39. F. Haber and J. Weiss: *Naturwiss.*, **20**, 948 (1932).
40. M. S. Kharash, F. S. Arimoto, and W. Nudenberg: *J. Org. Chem.*, **16**, 1556 (1951).
41. A. V. Tobolsky, D. J. Metz, and R. B. Mesrobian: *J. Amer. Chem. Soc.*, **72**, 1942 (1950).
42. Z. G. Kozlova, V. F. Tsepalov, and V. Ya. Shlyapintokh: *Kinet. katal.*, **5**, 868 (1964).
43. I. V. Zakharov and V. Ya. Shlyapintokh: *Kinet. katal.*, **4**, 706 (1963).
44. I. P. Skibida: *Ups. Khim.*, **54**, 1729 (1975).
45. R. A. Sheldon and J. K. Kochi: *Advan. Catal.*, **25**, 272 (1976).
46. Y. Kamiya and M. Kashima: *J. Catal.*, **25**, 326 (1972).
47. D. A. S. Ravens: *Trans. Faraday Soc.*, **55**, 1768 (1959).
48. C. E. H. Bawn and T. K. Wright: *Discuss. Faraday Soc.*, **46**, 164 (1969).
49. Y. Kamiya: *J. Catal.*, **33**, 480 (1974).
50. I. V. Zakharov: *Kinet. katal.*, **15**, 1457 (1974).
51. I. V. Zakharov and Yu. V. Geletii: *Neftekhimiya*, **18**, 261 (1978);
     idem: ibid, 18, 615 (1978).
52. Yu. V. Geletii and I. V. Zakharov: *Kinet. katal.*, **22**, 261 (1981).
53. A. Onopchenko and J. G. D. Schulz: *J. Org. Chem.*, **38**, 909 (1973);
     idem: ibid, **38**, 3727 (1973).
54. K. Tanaka: *Chemtech.*, 555 (1974).
55. O. Hayaishi: *Molecular Mechanisms of Oxygen Activation*, p. 1, Academic Press, New York, London (1974).
56. A. Bach and R. Chodat: *Ber. Deut. Chem. Ges.*, **36**, 600 (1903).
57. O. Warburg: *Heavy Metal Prosthetic Groups and Enzyme Actions*, Oxford Univ. Press, London and New York (1949).
58. H. Wieland: *On the Mechanism of Oxidation*, Yale Univ. Press, New Haven, Connecticut (1932).
59. M. T. Abbott and S. Udenfriend: *Molecular Mechanisms of Oxygen Activation* (Ed. O. Hayaishi), p. 167, Academic Press, New York and London (1974).
60. A. I. Archakov: *Mikrosomalnye okisleniya* (Microsomal Oxidation), Nauka, Moscow (1975).
61. M. Hamberg, B. Samuelsson, I. Björkhem, and H. Danielsson: *Molecular Mechanisms of Oxygen Activation* (Ed. O. Hayaishi), p. 29, Academic Press, New York and London (1974).
62. I. Björkhem: *Bioch. Bioph. Res. Commun.*, **50**, 581 (1973).
63. U. Frommer and V. Ullrich: *Z. Naturforsch.*, **26b**, 322 (1971).
64. U. Frommer, V. Ullrich, H. Staudinger: *Hoppe-Seyler's Z. Physiol Chem.*, **351**, 903 (1970);
     idem: ibid, 351, 913 (1970).
65. U. Frommer, V. Ullrich, and S. Orrenius: *FEBS Letts*, **41**, 14 (1974).
66. V. Ullrich: *Z. Naturforsch.*, **24b**, 699 (1969).
67. S. Udenfriend, J. W. Daly, G. Guroff, D. Gerina, P. Salbzma-Mirenberg, and B. Witkop: *Microsomes and Drug Oxidation* (Eds., J. R. Gilette *et al.*), p. 225, No. 1, Academic Press, New York and London (1969).

68. H. W. Strobel, A. Y. H. Lu, J. Heidema, and M. J. Coon: *J. Biol. Chem.*, **245**, 485 (1970).
69. R. W. Estabrook, A. G. Hildebrant, J. Baron, K. J. Netter, and K. Leibman: *Bioch. Biophys. Res. Commun.*, **42**, 132 (1971).
70. V. Ullrich: *J. Molec. Catal.*, 7, 159 (1980).
71. G. I. Likhtenstein: *Mnogojadernye okislitelno-vosstanovitelnye fermenty* (Multinuclear Redox Enzymes), Nauka, Moscow (1979).
72. V. Ullrich: *The Mechanism of Cytochrome P-450 Action*, p. 192 (Microsomes and Drug Oxidation), Pergamon Press, Oxford (1977).
73. G. A. Hamilton: *J. Amer. Chem. Soc.*, 86, 3391 (1964).
74. G. A. Hamilton: *Advan. Enzymol.*, 32, 55 (1969).
75. V. Ullrich and Hj. Staudinger' *Biochemie des Squerstoffs* (Eds. Hj. Staudinger and Hess), p. 229, Springer Verlag, Berlin-Heidelberg-New York (1968).
76. J. A. Peterson, Y. Ishimura, J. Baron, and R. W. Estabrook: *Oxidases and Related Redox Systems* (Eds. T. E. King, H. S. Mason, and M. Morrison), University Park Press, Baltimore, p. 565 (1973).
    J. T. Groves and G. A. McClusky: *J. Amer. Chem. Soc.*, 98, 859 (1976).
77. F. Lichtenberger, W. Nastainczyk and V. Ullrich: *Bioch. Biophys. Res. Commun.*, 70, 939 (1976).
78. D. Dolphin, A. Forman, D. C. Borg, J. Fajer, and R. H. Felton: *Proc. Natl. Acad. Sci. USA*, 68, 614 (1971).
79. D. Dolphin, R. H. Felton: *Acc. Chem. Res.*, 7, 26 (1974).
80. S. G. Sligar, K. A. Kennedy, and D. C. Pearson: *Proc. Natl. Acad. Sci. USA*, 77, 1240 (1980).
81. H. Dalton: *Adv. Appl. Microbiol.*, 26, 71 (1980).
82. I. J. Higgins, D. J. Best, and R. C. Hammond: *Nature*, 286, 561 (1980).
83. G. M. Tonge, D. E. F. Harrison, and I. J. Higgins: *Biochem. J.*, 161, 333 (1977).
84. J. Colby, D. I. Stirling, and H. Dalton: *Biochem. J.*, 165, 395 (1977).
85. H. Dalton, B. T. Golding, B. W. Waters, R. Higgins, and J. A. Taylor: *J. Chem. Soc., Chem. Commun.*, 482 (1981).
86. J. Colby and H. Dalton: *Biochem. J.*, 171, 461 (1978).
87. A. I. Pilyashenko-Novokhatnyi, A. N. Grigorjan, A. P. Kovalev, V. S. Belova, and R. I. Gvozdev: *Dokl. Akad. Nauk SSSR*, 245, 1501 (1979).
88. C. F. Schönbein: *Ber. natur. Ges. Bas.*, VI, 16 (1844).
89. C. F. Schönbein: *J. für praktische Chemie*, 55, 1 (1852).
90. N. A. Shilov: *O sopryazhennykh reaktsijakh okislenija* (On the Coupled Oxidation Reactions) Moscow (1905).
91. W. Manchot and O. Wilhelms: *Ber.*, 34, 2479 (1901).
91. W. Manchot: *Ann.*, 325, 93 (1902).
93. S. Udenfriend, C. T. Clark, J. Axelrod, and B. B. Brodie: *J. Biol. Chem.*, 208, 731 (1954).
94. V. Ullrich and Hj. Staudinger: *Z. Naturforsch.*, 24b, 583 (1969).
95. A. A. Akhrem, D. I. Metelitsa, and M. E. Skurko: *Usp. Khim.*, 44, 868 (1975).
96. N. Z. Muradov, A. E. Shilov, and A. A. Shteinman: *Kinet. katal.*, 13, 1357 (1972).
97. E. I. Karasevich, N. Z. Muradov, and A. A. Shteinman: *Izv. Akad. Nauk SSSR, ser. khim.*, No. 8, 1805 (1974).
98. N. Z. Muradov and A. A. Shteinman: *Izv. Akad. Nauk SSSR, ser. khim.*, No. 10, 2294 (1975).
99. E. T. Denisov: *Liquid-Phase Reaction Rate Constants*, p. 771, Plenum Press, New York, Washington, London (1974).

100. E. A. Kutner and B. P. Matseevskii: *Kinet. katal.*, **10**, 997 (1969).

101. H. Mimoun and I. Serée de Roch: *Tetrahedron*, **31**, 777 (1975).

102. A. P. Moravskii: *Thesis*, Moscow (1975).

103. I. V. Zakharov, E. I. Karasevich, A. E. Shilov, and A. A. Shteinman: *Kinet. katal.*, **16**, 1151 (1975).

104. Yu. V. Geletii, I. V. Zakharov, E. I. Karasevich, and A. A. Shteinman: *Kinet. katal.*, **20**, 1124 (1979).

105. E. I. Karasevich: *Thesis*, Moscow (1980).

106. G. Henrici-Olivé and S. Olivé: *Coordination and Catalysis*, Verlag Chemie, Weinheim-New York (1977).

107. J. P. Collman, R. R. Gagné, C. A. Reed, T. R. Halbert, G. Lang, and W. T. Robinson: *J. Amer. Chem. Soc.*, **97**, 1427 (1975).

108. J. P. Collman, R. R. Gagné, H. B. Gray, and J. W. Hare: *J. Amer. Chem. Soc.*, **96**, 6522 (1974).

109. a. M. F. Budyka: *Dokl. Akad. Nauk SSSR*, **257**, 115 (1981);
     b. M. F. Budyka, A. M. Khenkin, and A. A. Shteinman: *Bioch. Bioph. Res. Commun.*, **101**, 615 (1982);
     c. M. Shappacher, L. Ricard, R. Weiss, U. Gonsel, and A. Trautwein: *J. Amer. Chem. Soc.*, **103**, 7646 (1981);
     d. C. K. Chang, and D. Dolphin: *Proc. Natl. Acad. Sci. USA*, **73**, 3338 (1976).

110. a. I. B. Afanasiev, S. V. Prigoda, A. M. Khenkin, and A. A. Shteinman: *Dokl. Akad. Nauk SSSR*, **236**, 641 (1977);
     b. E. McCandish, A. R. Miksztal, M. Nappa, A. Q. Sprenger, J. S. Valentine, J. D. Stone, and T. G. Spiro: *J. Amer. Chem. Soc.*, **102**, 4268 (1980);
     c. A. M. Khenkin and A. A. Shteinman: *Kinet. Katal.*, **23**, 219 (1982).

111. V. M. Berdnikov, L. L. Makarshin, and L. S. Ryvkina: *React. Kinet. Catal. Lett.*, **9**, 281 (1978).

112. A. A. Akhrem, D. I. Metelitsa, and M. E. Skurko: *Bioorganic Chem.*, **4**, 307 (1975).

113. J. T. Groves and M. Van der Puy: *J. Amer. Chem. Soc.*, **98**, 5290 (1976).

114. N. Z. Muradov, A. E. Shilov, and A. A. Shteinman: *React. Kinet. Catal. Lett.*, **2**, 417 (1975).

115. J. E. Lyons: *Aspects of Homogeneous Catalysis* Vol. 3 (Ed. R. Ugo), p. 3, D. Reidel, Dordrecht, Boston (1977).

116. N. Z. Muradov: *Thesis*, Chernogolovka (1975).

117. V. N. Glushakova, N. A. Skorodumova, I. V. Spirina, and G. A. Rasuvaev: *XII Colloquium on Organometallic Chem.*, G. D. R.-Poland, Potsdam (1980).

118. V. P. Maslennikov, V. N. Alyasov, G. I. Makin, and Yu. A. Aleksandrov: *Zhur. obsch. khim.*, **46**, 82 (1976).

119. I. V. Spirina, V. P. Sergeeva, V. P. Maslennikov, N. N. Vyshinsky, V. N. Kokorev, and Yu. A. Aleksandrov: *Zhur. obshch. khim.*, **49**, 2509 (1979).

120. I. V. Spirina, V. P. Sergeeva, V. P. Maslennikov, V. G. Tsvetkov, and Yu. A. Aleksandrov: *Zhur. obshch. khim.*, **50**, 1826 (1980).

121. K. B. Sharpless and T. C. Flood: *J. Amer. Chem. Soc.*, **93**, 2316 (1971).

122. J. T. Groves, T. E. Nemo, and R. S. Myers: *J. Amer. Chem. Soc.*, **101**, 1032 (1979).

123. J. T. Groves, W. J. Kruper, Jr., T. E. Nemo, and R. S. Myers: *J. Molec. Catal.*, **7**, 169 (1980).

124. C. K. Chang and Ming-Shang Kuo: *J. Amer. Chem. Soc.*, **101**, 3413 (1979).

125. J. T. Groves, W. J. Kruper, Jr., and R. C. Haushalter: *J. Amer. Chem. Soc.*, **102**, 6375 (1980).

126. J. T. Groves, R. C. Haushalter, M. Nakamura, T. E. Nemo, and B. J. Evans: *J. Amer. Chem. Soc.*, **103**, 2844 (1981).

127. A. M. Khenkin and A. A. Shteinman: *Izv. Akad. Nauk SSSR, ser. khim.*, 1668 (1982).
128. J. L. Franklin: *Trans. Faraday Soc.*, 443 (1952).
129. A. F. Shestakov and A. E. Shilov: *Khimich. Phizika*, in press.
130. C. Naccache, J. Bandiera, and M. Dufaux: *J. Catal.*, 25, 334 (1972).
131. F. S. Dyachkovskii and A. E. Shilov: *Usp. khimii*, 35, 699 (1966).

CHAPTER V

ACTIVATION OF ALKANES BY METAL COMPLEXES OF MEDIUM AND LOW OXIDATION STATE

V.1. General Remarks

By the end of the 1960s a large quantity of data had accumulated, indicating the possibility of activating alkanes by transition metal complexes, when the alkane molecule reacts directly with the metal center by entering into the coordination sphere of metal complexes to form intermediately alkyl derivatives. These data may be summarized as follows:

1. A large number of complexes react with various compounds, containing C—H bonds, particularly with aromatic hydrocarbons, with rupture of the C—H bond and the formation of a metal-carbon fragment. Thus, the strength of the C—H bond, which is no lower in arenes than in alkanes, does not prevent a reaction.

2. An aliphatic C—H bond can be activated if it is present in a ligand attached to a metal center in the complex. It is an indication of the possibility (at least under more severe conditions) of analogous alkanes reactions.

3. Reactions of alkanes on the surface of metals or their compounds which undoubtedly involve the C—H bond cleavage in an alkane molecule also indicate the possibility of such reactions with metal complexes in homogeneous solutions, at least for polynuclear cluster complexes containing several metal atoms.

4. The reactions of alkanes in homogeneous solutions with electrophilic oxidizing agents containing metal in a high oxidation state sometimes suggest the possibility of direct participation of the metal ion in the reaction with the alkane to form a metal-carbon bond, although, as we have seen, in no case has any direct evidence for this been obtained.

As was pointed out, starting from analogies with the activation of other compounds with no π-bonds present — in particular dihydrogen — it may be assumed that the initial alkane reaction with compounds of transition metals in a low oxidation state will be an oxidative addition

$$M + RH \rightleftharpoons M\underset{H}{\overset{R}{\diagdown}} \tag{V.1}$$

In view of the fact that the cleavage of the strong C—H bond may require a considerable activation energy, the necessity of studying reactions of this type at elevated temperatures was predictable. However, it was necessary to take into consideration the possibility of thermodynamic instability of the alkyl hydride formed, i.e. that the reaction of oxidative addition might be displaced to the left, resulting in an equilibrium concentration of alkyl hydride too low to be directly detected, for example, by spectral methods. Hence it was necessary to find an indicator that would show the existence of a reversible reaction of Type V.1. A possible indicator is, for example, H—D exchange which has been successfully applied to the study of heterogeneous catalytic reactions and also different homogeneous reactions, e.g., those of dihydrogen and arenes.

Two types of reactions of H—D exchange can be used as indicators for the reversible Reaction V.1 [1, 2]. First, the alkyl hydride formed may have acidic properties and dissociate appreciably in protic media

$$M\begin{matrix} R \\ R \end{matrix} \rightleftharpoons [M-R]^- + H^+$$

In this case one can expect an exchange with solvent protons in protic media. In a solvent containing exchangeable deuterium, such as D_2O, H—D exchange must occur, serving as an indicator of the hydrocarbon-to-metal complex interaction.

Secondly, if the complex reacting with the alkane already contains hydride bonds, then an isotope exchange with molecular deuterium may be expected according to Scheme (V.1).

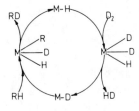

Scheme V.1

An attempt was naturally made to use the simplest complexes stable in protic media as catalysts for H—D alkane exchange with water. The obvious candidate for the role of such a catalyst was a very well known complex of bivalent platinum, $PtCl_4^{2-}$. It was known to catalyze the H—D exchange of D_2O with aromatic compounds [3, 4]. Moreover, the exchange had been shown to occur not only in the aromatic ring but also in the side chain of alkylbenzenes, including those C—H bonds which are not chemically activated, for example, the CH_3 group of ethylbenzene [3, 4].

The exchange was attributed to the π-complex formation between Pt(II) and the aromatic ring. However, a simple alternative was to assume a direct bimolecular interaction of platinum complexes with different C—H bonds of the alkylbenzene molecule. In this case alkanes must also be sufficiently reactive.

In a study set up in 1969 in this author's laboratory to investigate the possibility of alkane activation, it was demonstrated that this is precisely the case ([1], see also reviews [2]). Methane and ethane, when heated in sealed ampules containing solutions of K_2PtCl_4 in a mixture of D_2O—CH_3COOD, were found to be capable of exchanging hydrogen atoms with deuterons in the medium. Thus the activation of alkanes in solution was observed for the first time in the presence of a very simple complex which had been well-known for a long time. Let us note that this result could have been obtained much earlier, at least as early as the time when the fact of alkane H—D exchange with molecular deuterium on the surface of metals and their compounds was established. Apparently, the generally acknowledged inertness of alkanes created some kind of psychological barrier to the setting up of simple experiments to check the possibility of H—D exchange of alkanes with D_2O in the presence of ordinary Pt(II) complexes.

A positive result was also obtained in 1969 in the search for the second mechanism for the example of methane H—D exchange with D_2 in the presence of $(Ph_3P)_3CoH_3$, which was considered to be capable of reversible decomposition with the formation of the monohydride and H_2

$$(Ph_3P)_3CoH_3 \;\rightleftharpoons\; (Ph_3P)_3CoH + H_2$$

Monodeuteromethane was shown to be formed even at room temperature, the concentration of CH_3D rising with time [1].

Later on many results concerning the activation of alkanes by transition metal complexes in solution were published. The reactions of alkanes in the presence of Pt(II) were studied in particular detail. Although direct evidence for the formation of alkylhydride complexes in reactions with Pt(II) had not been obtained, it was unambiguously demonstrated that the exchange mechanism included the formation of platinum alkyl derivatives. The possibility of oxidative reactions of alkanes (as well as arenes) was discovered in the presence of a mixture of Pt(II)—Pt(IV) chloride complexes. It was demonstrated also that both the reactions (H—D exchange and oxidation) proceeded via the same alkyl (or aryl) platinum derivatives. In some cases aryl and alkyl complexes of platinum(IV) were isolated and characterized by means of different methods, X-ray structural analysis included.

Among the complexes catalyzing various reactions of alkanes, complexes of other metals of the platinum group (iridium, rhodium) were later included. Direct evidence for oxidative addition of alkanes to some iridium(I) and rhodium(I) complexes was obtained recently with detection of alkylhydride complexes.

Non-platinum metal complexes — those of titanium, vanadium, iron and rhenium — were found to be active towards alkanes in solution. New reactions previously unknown for alkanes were detected, such as the addition of CH_3—H to triple or double bonds of acetylenes and olefins.

Although the field of low-temperature alkane reactions involving metal complexes of medium and low oxidation state has not yet reached an importance comparable to that of catalytic hydrogenation, there is no doubt that this is a promising area of investigation which will continue to develop and should lead to new and important results.

V.2. Reactions of Arenes and Alkanes in the Presence of Platinum(II) Complexes

V.2.1. H–D EXCHANGE WITH THE SOLVENT

As has already been mentioned, the exchange of aromatic hydrocarbons with D_2O was discovered and studied by Garnett and coworkers [3, 4]. The isotope exchange of alkanes in the same system was discovered in 1969 [1] and investigated by several groups [2]. In many respects the behaviour of arenes and alkanes happened to be unexpectedly similar to each other, which is evidently the reflection of a similar mechanism. Hence, we shall review the results obtained for both types of hydrocarbons together, taking into consideration the peculiar behaviour of aromatic molecules among the other effects exerted by the nature of organic molecules on the reactivity of C—H bonds.

H–D exchange is usually observed at temperatures of 80–120°C. It takes place with hydrocarbons of the methane homologous series, their various derivatives (alkyl halides, carboxylic acids) and unsubstituted and substituted aromatic molecules. Bearing in mind that Pt(II) solutions can give precipitates of metallic platinum on heating and, particularly, in the presence of reducing agents, and also that metallic platinum is a catalyst for various reactions, including H–D exchange, the suspicion may arise that the exchange reaction actually proceeds by a heterogeneous catalytic mechanism on the surface of metallic platinum present, e.g., as colloid particles.

However, a number of experimental results leaves no doubt that the H–D exchange is actually a homogeneous reaction. The addition of mineral acids stabilizes Pt(II) solutions by making the following disproportionation reaction difficult

$$2\, PtCl_4^{2-} \;\rightleftharpoons\; Pt + PtCl_6^{2-} + 2\, Cl^- \tag{V.2}$$

but does not decrease the rate of exchange; on the contrary, precisely under those conditions, where $PtCl_4^{2-}$ solutions remain homogeneous, reproducible results are obtained. Molecular oxygen and the addition of aromatic compounds (e.g., pyrene) also stabilize homogeneous Pt(II) solutions and make the results more reproducible, preventing the formation of metallic platinum, which apparently catalyzes Reaction V.2.

On the other hand, on the precipitation of metallic platinum, the exchange slows down or stops completely. It must be noted that no metallic platinum is precipitated at all in the $PtCl_4^{2-}$–$PtCl_6^{2-}$ system, which oxidizes hydrocarbons at least in the early stages where sufficient amounts of Pt(IV) are present. As will be shown later, both H–D exchange and oxidation involve the same initial step of a reaction of Pt(II) with the hydrocarbon. Thus, there is no doubt at this time that homogeneous Pt(II) solutions react with hydrocarbons.

A 50% aqueous solution of acetic acid is very convenient to use as a solvent. K_2PtCl_4 solutions in aqueous acetic acid are more stable upon heating than aqueous solutions. Furthermore, in aqueous solutions, the solubility of hydrocarbons is rather low. Besides, even with a consideration of the different solubility of the hydrocarbons, the rate of alkane isotope exchange with D_2O in 50% acetic acid turned out to be about 30 times greater than in water. Since a methyl group of acetic acid also exchanges with D_2O in the presence of Pt(II), a mixture of $D_2O + CD_3CO_2D$ is used. The activating role of acetic acid can be attributed to the formation of a weak chelate complex

The ability of the acetic ion to act as a bidentate ligand is known. For example, it has been demonstrated in the investigation of the molecular structure of the complex $RuH(CH_3CO_2)(PPh_3)_3$ [5]. The H–D exchange rate is lower in trifluoroacetic acid, which has a lower chelating capacity than acetic acid.

By using H–D exchange catalyzed by $PtCl_4^{2-}$, we can obtain practically completely deuterated hydrocarbons if the reaction is carried out for a sufficiently long time. However, even in the initial stages of the reaction, di-, tri-, and polydeuterium-substituted molecules are formed. Thus several hydrogen atoms can be exchanged during a single contact between the hydrocarbon molecule and the platinum complex, i.e. a so-called multiple exchange takes place.

V.2.2. KINETICS OF H–D EXCHANGE

Important information on the mechanism of H–D exchange has been obtained in a study of reaction kinetics for both aromatic hydrocarbons and alkanes.

For the case of alkanes [6] in the presence of $HClO_4$, the H—D exchange rate within the $HClO_4$ concentration range of 0.2–1 M was shown to be virtually independent of the acidity and ionic strength. With $HClO_4$ concentrations lower than 0.2 M, the reaction rate decreases slightly.

The kinetics of the exchange reaction is first order with respect to hydrocarbon concentration, and of fractional order (< 1) with respect to the K_2PtCl_4 concentration. The dependence of the reaction rate on Cl^- ion concentration over a broad range of concentrations is shown for the case of cyclohexane in Fig. V.1. As the Cl^- concentration increases, the order with respect to the chloride-ion concentration changes from 0 to −1. This dependence permits

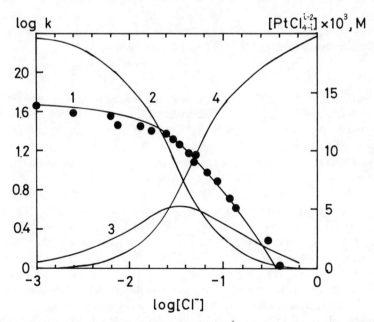

Fig. V.1.: The dependence of $\log k$ on $\log [Cl^-]$. Dots are the experimental values of $\log k$, the line is an estimation by formula V.3; (2), (3), (4) are equilibrium concentrations of $PtCl_2$, $PtCl_3^-$, and $PtCl_4^{2-}$, respectively.

us to draw conclusions on the nature of the species reacting with the alkane molecule. It is known that, in aqueous solution, $PtCl_4^{2-}$ ions undergo an equilibrium dissociation

$$PtCl_4^{2-} \underset{}{\overset{K_1(S)}{\rightleftharpoons}} SPtCl_3^- + Cl^-$$

$$SPtCl_3^- \underset{}{\overset{K_2(S)}{\rightleftharpoons}} S_2PtCl_2 + Cl^-$$

where S is the solvent.

The negative dependence on the Cl^- concentration shows that the species produced by $PtCl_4^{2-}$ dissociation are more active than $PtCl_4^{2-}$ itself.

The dependence of the $SPtCl_3^-$ and S_2PtCl_2 concentration on the Cl^- ions concentration is shown in Fig. V.1. We can see that the dependence of the rate corresponds approximately to the change in the concentration of uncharged species (S_2PtCl_2) in the system. If for S_2PtCl_2, $SPtCl_3^-$, and $PtCl_4^{2-}$ the rates constants are, respectively, k_1, k_2, and k_3, then the dependence of the effective rate constant (k) on the Cl^- concentration may be expressed by

$$k = \frac{k_1 + k_2 K_2^{-1}[Cl^-] + k_3 K_1^{-1} K_2^{-1}[Cl^-]^2}{1 + K_2^{-1}[Cl^-] + K_1^{-1} K_2^{-1}[Cl^-]^2} \tag{V.3}$$

which is in good agreement with experimental results (see Fig. V.1). An analysis of this dependence results in the following ratio of constants:

$$k_1 : k_2 : k_3 = 100 : 14 : 0.6$$

(for cyclohexane at $100°C$), i.e., in effect, the uncharged complex S_2PtCl_2 is the most active. This may be an indication of the importance of electrophilic properties of the platinum complex in the reaction with hydrocarbons, although an appreciable value of the rate constant of such a moiety as $PtCl_3^-$, in its reaction with alkanes in aqueous solution, does not very well fit the usual concepts of electrophiles.

Moreover, it has been shown [7] that the particular Pt(II) complexes obtained, which should form positive ions upon solution in water, $[PtCl(H_2O)_3]^+$ and $[Pt(H_2O)_4]^{2+}$, react more slowly with alkanes than $PtCl_2$, which demonstrates a more complicated nature of interaction of bivalent platinum and hydrocarbon molecule than that of the interaction of an electrophile with a nucleophile.

This conclusion is supported by the study of the effect of the other ligands on the H–D exchange rate [8]. In particular, the effect of the ligand's nature in the complexes $PtLCl_3^-$ ($L = H_2O$, Cl^-, NO_2^-, DMSO, NH_3, py), $PtCl_2L_2$ ($L = H_2O$, Cl^-, Br^-, I^-, NO_2^-, CN^-, PPh_3), and others, has been investigated. As can be seen from Table V.1, depending on the nature of the ligand, the rate

TABLE V.1

H–D exchange of cyclohexane in the presence of Pt(II) complexes.
$[Pt^{II}]$ = 0.0043 M (1, 2); 0.02 M (3–12); $[HClO_4]$ = 0.1 M;
$[C_6H_{12}]$ = 0.15 M; t = 100°C.

Nos.	Complexes	Time h	$\Sigma id_i{}^a$	k 10^3 M^{-1} s^{-1}	$M = \dfrac{\Sigma id_i{}^b}{\Sigma d_i}$
1	$PtCl_2S_2$	2	0.230	7.42	1.93
2	$KPtCl_3S$	2	0.154	4.97	1.74
3	K_2PtCl_4	1.66	0.333	2.78	1.75
4	$K_2Pt(NO_2)Cl_3$	7	0.185	0.366	1.52
5	$KPt(NH_3)Cl_3$	2.5	0.140	0.777	1.65
6	$KPt(py)Cl_3$	12	0.0268	0.025	1.46
7	$KPt(DMSO)Cl_3$	12.5	0.0549	0.056	1.60
8	$Pt(DMSO)Cl(NO_3)$	4	0.0242	0.083	1.62
9	$K_2Pt(NO_2)_2Cl_2$	7	0.0584	0.115	1.41
10	$Pt(PPh_3)_2Cl_2$	16.5	0.0078	0.007	1.34
11	$Pt(acac)_2$	7.5	0.0511	0.094	1.57
12[c]	$Pt(CF_3CO_2)_2$	0.3	0.0158	0.732	1.34

[a] d_i is the content (as a part of a whole) of deuterated hydrocarbon with i D atoms;
[b] $M = \Sigma id_i/\Sigma d_i$ (in the last column) is the multiple exchange factor equal to an average number of D atoms introduced in the hydrocarbon molecule during a single contact with the catalyst;
[c] Experiment carried with CF_3COOH at 91°C.

constant changes by three orders of magnitude, decreasing as a result of the introduction of more basic and also of more highly polarizable ligands and of ligands with an increased tendency to double bond formation.

Table V.2 lists the rate constants in the case of H–D exchange catalysis by the complexes formed in the system

$$PtCl_2S_2 + 2 X^- \rightleftharpoons PtCl_2X_2^{2-} + 2 S$$

The rate of H–D exchange decreases in the following order:

$$X = F > SO_4^- > Cl > Br > I > NO_2 > CN$$

which correlates with the change of PtX_4^{2-} complex stability constant and also with the overlap integral of the $6p\sigma$-orbital of platinum and the $p\sigma$-orbital of the ligand [9] and the parameter (σ_x) of the ligands [10], which characterizes their 'softness'.

The ligand will obviously effect both the concentration of PtX_2 complex in the solution and the electronic properties of the complex. The effect of the

TABLE V.2

Anion effect on H–D exchange of cyclohexane in the presence of $PtCl_2S_2 + 2\ X^-$;
Solv.: 50% aq. HOAC; [Pt(II)] = 0.0043 M; $[X^-]$ = 0.0086 M; $[HClO_4]$ = 0.1 M;
[pyrene] = 0.013 M; $[C_6H_{12}]$ = 0.15 M.

X^-	None	CF_3COO^-	F^-	SO_4^{2-}	Cl^-	Br^-	I^-	NO_2^-	CN^-
$k \times 10^3$ $M^{-1}\ s^{-1}$	6.30	6.28	6.26	6.06	4.10	2.47	0.394	0.115	0.106
σ_X	13.05		12.18		9.92	9.82	8.31	5.87	
M^a	2.02	1.94	1.92	1.98	1.77	1.48	1.52	1.41	1.52

a Multiple exchange factor.

ligand nature on the catalytic properties of the complex may be followed by a comparison of the activity of complexes of the type $PtCl_3L$ where $L = Cl^-$, H_2O, NH_3, and DMSO. The stability constants of the aquo-complexes and DMSO-platinum complexes are similar, i.e. the concentrations of the active complexes $PtCl_2$–(DMSO) and $PtCl_2(H_2O)S$ in solution are close, whereas the rate constants of H–D exchange with cyclohexane for those complexes differ by a factor of over ten. Dimethyl sulfoxide, which is a strong π-acceptor, should give rise to a higher positive charge on the central ion which, in agreement with the concept of the electrophilic character of its interaction with the hydro-carbon, should have resulted in a higher rate of H–D exchange (as in the re-placement of NH_3 by H_2O). The slower rate of H–D exchange in the case of DMSO, as compared with H_2O, which is actually observed, is apparently connected with an increase (in comparison with H_2O) of the softness of the sulfur-containing ligand. Soft ligands are well-known to stabilize Pt(II) com-plexes against oxidation to Pt(IV). All the ligands tested can be arranged in the following order, reflecting their influence on the rate of deuterium exchange:

$$PPh_3 \simeq py < DMSO < CN^- < NO_2^- < NH_3 < I^- < Br^- < Cl^- < F^- \simeq H_2O$$

It is interesting to note that this order is opposite to the known order for the *trans*-effect of ligands in substitution reactions of square platinum complexes. Therefore, the result obtained does not agree with the mechanism of usual electrophilic H^+ substitution in hydrocarbons, where the hydrocarbon molecule acts as a nucleophile.

$$RH + Pt^{II}-Cl \longrightarrow [H^{\delta+} \ldots R \ldots Pt \ldots Cl^{\delta-}] \longrightarrow H^+ + RPt + Cl^-$$

since a normal order of *trans*-effect would have been expected for a mechanism of this type. The order in the ligand effect is apparently determined by the strength of the covalent Pt—C bond formed and, possibly, of the Pt—H bond as well, both of which are weaker for softer ligands. Thus, the vibrational frequency of the Pt—C bond, which apparently changes approximately linearly with the energy of the bond dissociation, decreases in the order [11]

$$NO_3^-, NCS^- \gg Cl^- > Br^- > NO_2^- > I^- > CN^-$$

A rather similar order of ligand effects is observed for vibrational frequencies of the Pt—H bond in PtA_2LH complexes [12]:

$$NO_3^- > Cl^- > Br^- > I^- > NO_2^- > SCN^- > CN^-$$

In both the series the order is close to opposite to that of the ligand *trans*-effects in substitution.

The isotope effect, though moderate, observed in the reaction of H—D exchange, indicates the participation of the C—H bond cleavage in the rate-determining step. In the systems CD_4-H_2O and CH_4-D_2O, the ratio of the constants $k_{CH_4}/k_{CD_4} = 3.0 \pm 0.5$ (100°C) [6], for cyclohexane $k_{C_6H_{12}}/k_{C_6D_{12}} = 1.7 \pm 0.1$ (100°C) [13], for benzene $k_{C_6H_6}/k_{C_6D_6} = 1.65$ [4].

V.2.3. RELATIVE REACTIVITY OF HYDROCARBONS IN H–D EXCHANGE

Table V.3 lists the data for H—D exchange of the various hydrocarbons in the presence of $PtCl_4^{2-}$ reported by Hodges, Webster and Wells [13].

The data in the Table demonstrate first of all that *n*-alkanes are more reactive than branched alkanes.

In normal and branched alkanes the highest activity is observed for isolated methyl groups. For example, according to [13], in normal pentane the rate of isotope exchange in two methyl groups is 4.5 ± 0.8 times greater than the rate of exchange in the three methylene groups. In alkanes containing primary, secondary and tertiary C—H bonds the reactivity changes in the order $1° > 2° > 3°$ which is opposite to the 'normal' orders of reactivity (e.g., with respect to free radical reactions).

Both the methyl and methylene groups show a particularly low reactivity when they are bonded to a quaternary carbon atom (e.g., in 2,2-dimethylpropane and 2,2-dimethylbutane). All these results are most naturally explained by the important role played by steric factors. Primary alkyl metal species are known to be more stable than secondary and, in particular, tertiary derivatives presumably due to steric hindrance in the arrangement of bulky secondary and tertiary alkyl groups around the metal atom.

Arenes show higher reactivity than alkanes, but this difference is not so pronounced as in many other cases: benzene, which is the least reactive arene,

TABLE V.3

Deuteration of alkanes and alkylbenzenes at 120°C in the presence of $PtCl_4^{2-}$ in CD_3CO_2D. Solvent: 50 mol.% of CD_3CO_2D in D_2O. [Alkane] or [alkylbenzene] = 0.5 M. [Pyrene] = 0.05 M; [$DClO_4$] = 0.2 M.

Compound	$[PtCl_4^{2-}]$ M	Time h	% D (mass-spec)		% D (NMR)	
			Obs.*	Calc. % D	Position	% D
$\overset{1}{C}H_4$	0.04	95.0	25.0	97.3	1	25.0
$\overset{1}{C}H_3CH_3$	0.04	136.6	91.4	96.2	1	91.4
$\overset{1}{C}H_3\overset{2}{C}H_2CH_2CH_2CH_3$	0.04	136.6	74.6	92.8	1	92 ± 2
					2	57 ± 2
$\overset{1}{C}H_3\overset{2}{C}H\overset{3}{C}H_2\overset{4}{C}H_3$	0.04	136.6	68.8	92.8	1 and 4	82 ± 2
$\quad\ \mid$					2	9 ± 3
$\quad\ CH_3$					3	37 ± 2
$(\overset{1}{C}H_2)_6$	0.01	110.0	70.1	92.8	1	70.1
$\quad\ CH_3$					1	4.15*
$\overset{1}{C}H_3\overset{}{C}\!-\!\overset{2}{C}H_2\overset{3}{C}H_3$	0.02	91.0	18.2	90.2	2	2 ± 2
$\quad\ \mid\ CH_3$					3	70.3*
$Ph\overset{1}{C}H_2\overset{2}{C}H_2\overset{3}{C}H_2\overset{4}{C}H_2\overset{5}{C}H_2\overset{6}{C}H_3$	0.03	92.0	38.1	89.6	1	70 ± 2
					2	41 ± 2
					3, 4 and 5	8 ± 2
					6	39 ± 2
$\quad\ \overset{2}{C}H_3$					1	64 ± 2
$Ph\!-\!\overset{1}{C}\!-\!\overset{3}{C}H_2\!-\!\overset{4}{C}H_3$	0.03	92.0	30.0	89.6	2 and 3	2 ± 2
$\quad\ \mid\ CH_3$					4	53 ± 1

* Determined by mass-spectrum of ion fragments.

exchanges its H atoms only twice as fast as the most reactive of the alkanes, cyclohexane. In the case of substituted aromatic molecules the steric effect is again very pronounced. In an H—D exchange of monosubstituted benzenes, only *m*- and *p*-hydrogen atoms of the ring are exchanged (with apparently similar reactivity). Thus, in the exchange only molecules containing a maximum of three D atoms are formed from bromobenzene and chlorobenzene. *p*-Disubstituted benzenes do not exchange ring-hydrogen atoms at all. For example, in *p*-xylene H—D exchange occurs only in the side chain of the ring. Naphthalene exchanges hydrogen atoms only in the *β*-position [14].

The reactivity of C—H bonds in a side chain in the α-position to the benzene ring is higher than in other positions (which is connected apparently with a low value of the C—H bond energy). However, as demonstrated by n-pentylbenzene from Table V.3, it is almost the same for the CH_2 group next to phenyl as for the methyl group, where the energy of the C—H bond dissociation in the side chain is the highest.

The importance of the electron donor properties of hydrocarbons in reactions of H–D exchange is stressed by Hodges et al. [13], who reported a linear correlation between the logarithm of the exchange rate constant and the ionization potential of respective hydrocarbons (Fig. V.2). It is of interest that points corresponding to alkanes and aromatic hydrocarbons are located on the same

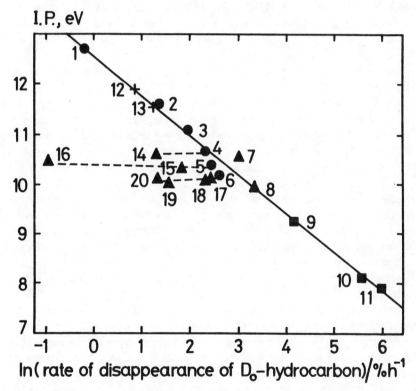

Fig. V.2: The dependence of the rate of H–D exchange on the ionization potentials of alkanes and aromatic hydrocarbons (1 – methane, 2 – ethane, 3 – propane, 4 – n-butane, 5 – n-pentane, 6 – n-hexane, 7 – cyclopentane, 8 – cyclohexane, 9 – benzene, 10 – naphthalene, 11 – phenanthrene, 12 – 2,2-dimethylbutane, 13 – 1,1-dimethylpropylbenzene, 14 – 2-methylpropane, 15 – 2-methylbutane, 16 – 2,2-dimethylpropane, 17 – 2-methylpentane, 18 – 3-methylpentane, 19 – 2,3-dimethylbutane, 20 – 2,2-dimethylbutane) [2b].

straight line, which is evidence for a similar activation mechanism for the two types of hydrocarbons.

However, branched hydrocarbons naturally do not show the above mentioned dependence, due to the steric effect discussed above. In the case of halogen-substituted alkanes the relation between the rate of exchange and the ionization potential is reported to have a different slope [15]. The reaction rates of cyclic hydrocarbons — cycloheptane and cyclooctane — also do not correlate with the ionization potentials. Obviously, there is no reason to expect a simple (e.g., linear) dependence of log k on the ionization potential for all hydrocarbons and their derivatives for the case of Pt(II)-catalyzed H—D exchange with a solvent. The rate determining step must involve both the C—H bond cleavage and the formation of the M—C bond, and, probably, of the M—H one. The facilitation of the reaction in the case of C—H bonds located in an α-position to the benzene ring as compared with simple alkanes is related to a decrease of dissociation energy of these bonds. At the same time, for aromatic hydrocarbons, higher rates of the reaction (as compared with alkanes) might be caused mainly by a more stable M—aryl bond as compared with the M—alkyl one.

All these effects are not expected to correlate with only one parameter, such as the ionization potential, even if this is considered as the potential of an electron of the orbital which is affected in the reaction with the platinum complexes. The apparent linearity of the observed log k relation with the ionization potential might be connected to a considerable extent with the fact that the effect of C—H bond differences on the rate of H—D exchange in case of Pt complexes is relatively small. When we go from the most inert alkanes of the methane series to the most active aromatic hydrocarbons (such as phenanthrene and pyrene), the rate differs by slightly more than two orders of magnitude, whereas the ionization potential changes by over 5 eV. This would have changed the reaction rate by a factor of approximately 10^{64} at 120°C if the latter were determined entirely by the energy of electron transfer from the hydrocarbon molecule.

Another approach to the quantitative correlation between the structure and the reactivity of hydrocarbons may be the use of the Taft correlation equation.

A consideration of the dependence of the exchange rate on the constant σ* for a number of functionally substituted methane and ethane derivatives has shown [16] that the two-parameter equation is obeyed (Fig. V.3)

$$\log k_s/k_0 = \rho^*\sigma^* + n\psi \tag{V.4}$$

Thus, besides the dependence on the polar parameter σ* (which characterizes the inductive effect of a substituent) the reaction rate depends also on the resonance term ψ which characterizes the 'conjugation' of the α-substituent

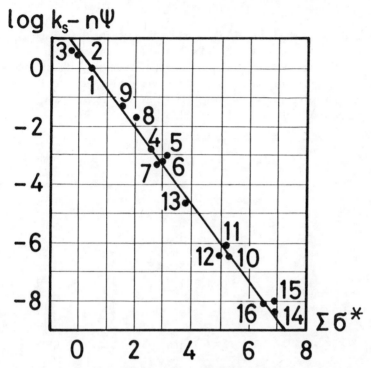

Fig. V.3: The dependence of the rate of H–D exchange on the Taft polar constant (1 – methane, 2 – ethane, 3 – propane, 4 – 1,1,1-trifluoroethane, 5 – fluoromethane, 6 – chloromethane, 7 – bromomethane, 8 – acetic acid, 9 – methanol, 10 – difluoromethane, 11 – dichloromethane, 12 – fluoromethane, 13 – 1,1-difluoroethane, 14 – trifluoromethane, 15 – trichloromethane, 16 – tribromomethane).

with the radical reaction center or with the π-system (n is equal to the number of such substituents); the value of ρ^* which has been found to be equal to -1.4, shows that Pt(II) acts as a moderate electron acceptor with respect to hydrocarbon, which acts as a donor in the reaction.

The presence of a resonance term in the correlation equation indicates the stabilizing effect of the electron pair of the substituent in the cleavage of the C—H bond in the interaction with the platinum complex, which could have taken place in the homolytic cleavage of the C—H bond to produce a quasi-free radical in the transition state.

Taking into consideration the importance of steric factors in the reaction, Equation V.4 should be supplemented naturally with one more term to account for steric hindrance in order to extend it to other hydrocarbons.

V.2.4. MULTIPLE EXCHANGE AND ITS MECHANISM

As already mentioned, the H—D alkane exchange with the solvent catalyzed by Pt(II) is multiple, i.e. the hydrocarbon molecule entering into the reaction may exchange several hydrogen atoms with deuterium without leaving the coordination sphere of a catalyst. Table V.4 presents the values of the parameter M which shows the average number of H atoms replaced by D during a contact of a hydrocarbon molecule with the catalyst [13].

TABLE V.4

Multiple exchange parameters of different hydrocarbons

Compounds	[RH] M	t °C	reaction time h	% D	M
Methane	0.3	100	25.50	7.80	1.48
Ethane	0.3	100	5.08	5.56	1.70
Propane	0.3	100	5.08	7.16	1.63
Butane	0.3	100	2.00	3.33	1.62
Pentane	0.3	100	4.00	6.39	1.65
	0.2	80	15.83	4.86	1.87
Hexane	0.3	100	5.08	6.65	1.50
Cyclopentane	0.3	100	2.00	6.45	1.68
Cyclohexane	0.3	100	2.00	6.31	1.42
	0.2	80	4.33	2.81	1.61
Benzene	0.2	80	4.00	21.18	3.50
Naphthalene	0.2	80	2.0	10.92	1.02
Phenanthrene	1.3	80	2.45	13.3	0.95

[Pt] = 0.02 M; Cl : Pt = 4 : 1; [DClO$_4$] = 0.2 M; 50 mole % AcOD/D$_2$O.

For alkanes this parameter has a value of 1.4–1.7, for benzene it increases to 3.5, and for naphthalene and phenanthrene it drops down to 1. The value of M increases only slightly with decreasing temperature.

The phenomenon of multiple exchange is well known for heterogeneous catalysis and is usually explained by the intermediate formation of surface π-complexes from alkyl metal derivatives (see Section III.1). A similar mechanism may be proposed for the homogeneous exchange of ethane and other hydrocarbons.

$$Pt(II) + C_nH_{2n+2} \underset{-H^+}{\rightleftharpoons} Pt-C_nH_{2n+1} \underset{-H^+}{\rightleftharpoons} Pt \cdots \overset{\diagup}{\underset{\diagdown}{\overset{C}{\underset{C}{\|}}}}$$

$$\mathbf{I} \qquad\qquad \mathbf{II} \qquad\qquad \mathbf{III}$$

Each I–II–I cycle amounts to the introduction of one D atom into the alkane molecule in the deuterated solvent ($[D^+] \gg [H^+]$). If the rates of conversion of II $\longrightarrow$ I and II $\longrightarrow$ III are comparable, then every cycle I $\longrightarrow$ II $\longrightarrow$ I may involve several cycles of II $\longrightarrow$ III $\longrightarrow$ II which leads to the insertion of several deuterium atoms during the contact of one alkane molecule with the Pt complex.

Multiple exchange in the case of methane may be accounted for in the same way assuming the intermediate formation of a methylene–platinum complex, Pt=CH$_2$. At present, a number of carbene complexes with different transition metals are known, including those of methylene CH$_2$, e.g., with tantalum. It is interesting that the methylene complex of tantalum is formed by deprotonation of a cationic methyl-tantalum complex [17].

$$\left[(C_5H_5)_2 Ta \overset{\diagup CH_3}{\diagdown CH_3} \right]^+ \xrightarrow{\; -H^+ \;} (C_5H_5)_2 Ta \overset{\diagup CH_2}{\diagdown CH_3}$$

i.e. in a reaction analogous to the one assumed for methane multiple exchange in the presence of platinum complexes.

The distribution of deuterohydrocarbons, which does not depend on time during the initial reaction stages, can be evaluated by a method of steady-state concentrations, for example, for H–D exchange in methane:

$$
\begin{array}{ccccc}
CH_4 & CH_3D & CH_2D_2 & CHD_3 & CD_4 \\
\searrow \; \nearrow{\scriptstyle k_2} & & \nearrow & \nearrow & \nearrow \\
Pt-CH_3 & & PtCH_2D & PtCHD_2 & PtCD_3 \quad (V.5) \\
\searrow{\scriptstyle k_3} & \nearrow\!\!\searrow & \nearrow\!\!\searrow & \nearrow\!\!\searrow & \nearrow \\
& Pt=CH_2 & Pt=CHD & Pt=CD_2 &
\end{array}
$$

The amount of each of the deuteromethane present in the mixture ($\alpha_i = d_i / \Sigma d_i$) is completely determined by the factor k_3/k_2. A calculation of deuterocarbon distribution (α_i) for methane and ethane in a $1:1$ CH$_3$CO$_2$D–D$_2$O mixture shows that Scheme V.5 adequately describes the experimental results [18].

The multiple exchange in alkanes, which is very sensitive to reaction conditions, may be used to prove the formation of alkyl platinum derivatives as intermediate products of the exchange. Alkyl platinum derivatives can be produced using organo-mercuric compounds via the exchange reaction

$$RHgBr + PtCl_2 \longrightarrow RPtCl + HgBrCl$$

The alkanes produced in this interaction are almost certainly formed as a result of a hydrolytic cleavage of alkyl platinum derivatives. Hence, the comparison of the distribution of deuteroalkanes formed under similar conditions in H–D exchange of alkanes with deuterated solvent in the presence of Pt(II) and in reaction of RHgBr with the same solvent also in the presence of Pt(II), if in both cases it is the same, will be an indication that deuteroalkanes are produced in both cases from the same source, that is from alkyl platinum complexes formed as intermediates. Table V.5 gives the distribution of deuteromethanes in the reaction of CH_3—HgBr with K_2PtCl_4 and for direct K_2PtCl_4-catalyzed H–D exchange in CH_4 under identical conditions ($[Pt^{II}]$: $[Hg]$: $[DCl]$ = 1 : 1 : 3) [19].

TABLE V.5

The distribution of deutero-methanes in the methylation of Pt(II) and direct H–D exchange

Conditions	$\alpha_i = \dfrac{d_i}{\Sigma d_i}$				D (%)	M
	α_1	α_2	α_3	α_4		
K_2PtCl_4 : $HgCl_2$: DCl = 1 : 1 3, P_{CH_4} = 760 Torr 100°C, 9 h	0.650	0.251	0.0735	0.0296	0.915	1.4
K_2PtCl_4 : CH_3HgBr : DCl = 1 : 1 : 3, 100°C, 0.3 h	0.630	0.268	0.0815	0.0212	35.0	1.4

The close agreement between the ratio of d-methanes formed in Pt(II) methylation and in direct H–D exchange with alkanes is a sound support for the formation of alkyl platinum derivatives in the homogeneous activation of saturated hydrocarbons.

A distribution of deuterohydrocarbons in the case of methane and ethane in a 1 : 1 mixture of CH_3CO_2D—D_2O is smoothly descending, i.e. the greater the number of deuterium atoms in the hydrocarbon, the lower will be the proportion of the corresponding hydrocarbon.

Experiments on the H–D exchange of ethane in pure D_2O have demonstrated, however, an unusual stepwise distribution of deuteroethanes which may be visualized in the form of two groups, $\alpha_1 - \alpha_3$ and $\alpha_4 - \alpha_6$, the distribution of each of these groups ascending, although the total content of ethane in the group $\alpha_4 - \alpha_6$ is considerably smaller than that in the group $\alpha_1 - \alpha_3$. It was shown by mass-spectrometry of the deuteroethanes that the asymmetrical molecules CHD_2CH_3, CD_3CH_3, CD_3CH_2D prevail in the $d_2 - d_4$ deuteroethanes. Such a distribution for the H–D exchange in water (predominance of $d_1 - d_3$ over

d_4-d_6 and the same ascending pattern in each group) suggests that the exchange takes place first in one of methyl groups of ethane and then in the other, which is thus not equivalent to the first. Obviously, such a distribution is inconsistent with the mechanism involving an intermediate π-complex formation in which both carbon atoms should be equivalent. There is a logical explanation of a stepwise distribution if it is assumed that the multiple exchange in ethane is similar to that in methane involving the intermediate formation of complexes with carbene, $Pt\cdots CHCH_3$ formed as a result of α-elimination of a proton. In this case, not more than three atoms of one methyl group will initially be exchanged and then, provided there is an elementary act of rotation of the alkyl group to form a Pt—C bond with another carbon atom, exchange in another methyl group will produce polydeuterated molecules with up to all six atoms exchanged, i.e. C_2D_6. The isomerization will probably proceed via the intermediate formation of a π-complex:

$$Pt(II) + C_2H_6 \underset{k_2}{\overset{k_1}{\rightleftharpoons}} Pt-C_2H_5 \begin{array}{c} Pt \cdots \overset{\displaystyle\|}{\underset{\displaystyle C}{C}} \\ Pt=CHCH_3 \end{array} \qquad (V.6)$$

A calculation of the deuteroethane distribution by the method of steady-state concentrations according to Scheme V.6 is in good agreement with experimental data obtained for an aqueous solution.

It is interesting to note that the smoothly descending distribution in deuteroethanes (obtained for solutions in 50% acetic acid) may be explained by both mechanisms, the one involving the ethylidene complex and the other consistent with only π-complex formation. In effect, if we accept that the conversion $Pt-C_2H_5 \rightarrow Pt=CHCH_3$ occurs more slowly than the turnover of the alkyl group $PtCH_2^1CH_3^2 \rightleftharpoons PtCH_2^2CH_3^1$, the carbene mechanism and that with the participation of only π-complexes will give the same distribution. However, it is more natural to assume that the carbene derivative is present both in aqueous and in acetic acid solutions, and it is the ratio of rate constants that changes, than to assume that the mechanism changes completely with a change of the solvent composition.

In the case of other alkanes a stepwise distribution is also observed with a clearly distinguishable group $\alpha_1-\alpha_3$ (Fig. V.4), which is an indication of a predominant exchange occurring in one of the methyl groups. This shows that platinum is more likely to be bonded with one (terminal) carbon atom with a transition 1-(alkyl) → 1,1-(carbene) (α-elimination) than to form a 1,2-(olefin)-bonded fragment (β-elimination). The isomerization proceeding via

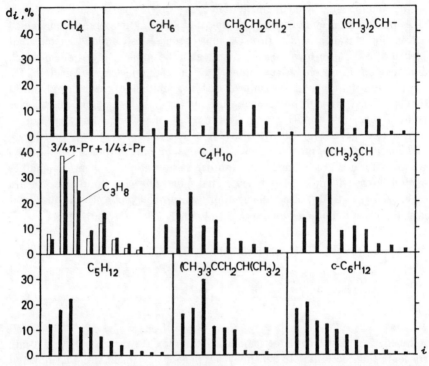

Fig. V.4. The distribution of deuteroalkanes according to their masses. $CH_3CH_3CH_2^-$, $(CH_3)_2CH$-distribution of deuteropropanes in the case of platinum alkylation with n-PrHgBr and i-PrHgBr, respectively.

1,2-bonded fragments (or π-complexes) leads to an extensive multiple exchange to produce polydeuterated alkanes via the mechanism $1,1 \overset{*}{\rightarrow} 1,2 \rightarrow 2,2$ or $1,1 \rightarrow 1 \rightarrow 1,2 \rightarrow 2,2$.

In the case of propane it is necessary to take into account the possibility of formation of the n- and i-propyl platinum complexes. An analysis of detueropropane distribution makes it possible to determine the relative rates of platinum(II) reactions with primary and secondary carbon atoms. For this purpose the distribution of deuteropropanes in the case of platinum alkylation with n-PrHgBr and i-PrHgBr was investigated. As shown in Fig. V.4, the distribution of the deuteropropanes obtained by the H–D exchange of propane coincides with that detected for n-PrPt- and i-PrPt-derivatives if we accept that they are initially formed with a relative probability of 0.75 and 0.25, respectively. Taking into consideration that the propane molecule contains six primary and two secondary C–H bonds, it may be concluded that the selectivity in the attack of primary and secondary bonds in propane is in accordance with the ratio $1° : 2° = 1 : 1$.

This demonstrates a much higher relative reactivity of a secondary C—H bond in propane as compared with the reactivity of secondary C—H bonds in pentane where, as previously noted, the probability of attack of a primary C—H bond is 4.5 times as great as that at a secondary bond. This can be caused by the smaller steric hindrances of the secondary C—H bond in propane bound to two methyl groups in comparison with secondary C—H bonds in pentane bound to, at least, one long alkyl group. Similarly, in linear alkanes, the CH_2 groups of the second carbon atom ($\omega - 1$ carbon) should show a higher reactivity than all the other CH_2 groups on analogy with some cases of biological oxidation described in Section IV.2. We shall see some evidence for this later. The presence of a second step $\alpha_4 - \alpha_5$ in the mass-spectrum of deuteropropanes in the platinum alkylation by n-PrHgBr is consistent with the 1,2-shift of the platinum along the carbon atoms; however, the considerable proportion of α_6 in the case of platinum alkylation using both n-PrHgBr and i-PrHgBr confirms the strong probability of a 1,3-shift of platinum. This shift may include a four-membered metallocycle with platinum participation [20]. Metallacycles of such a structure are known to be formed in the interaction of cyclopropane with K_2PtCl_4 [21].

These intermediate metallacycles are apparently formed in the multiple exchange of such hydrocarbons as isobutane and isooctane, for which a 1,2-shift should be sterically hindered. A 1,3-shift in the case of these hydrocarbons is supported by the triplet groups in the mass spectra of the corresponding hydrocarbons, and the considerable contribution to the distribution of the $\alpha_4 - \alpha_6$ group shows the ease of the 1,3-shift. The presence of a group of lines $\alpha_7 - \alpha_9$ for isobutane, and absence of an α_{10} line, in contrast to n-butane, shows that the tertiary C—H bond does not participate in the exchange in either the primary attack (as was already mentioned above) or in the 1,2-shift. The presence of a group of lines $\alpha_7 - \alpha_9$ in the isooctane exchange shows the possibility of a 1,4-shift. This result also demonstrates that an NMR-observed small exchange in CH_3 groups bonded to a quaternary carbon atom is at least partially due to a multiple exchange of the molecule, the rate of the initial Pt(II) interaction with these groups being even lower than follows from the NMR data.

The differences in the factor of multiple exchange for aromatic molecules (see Table V.4) could be explained to a considerable extent by steric factors. The large value of the factor M for benzene testifies for a high probability of the 1,2-shift (possibly via the intermediate formation of dehydrobenzene)

$$\text{Pt}-\!\!\overset{2}{\underset{}{\bigcirc}}\!\!-1 \xrightarrow{-\text{H}^+} \text{Pt} \leftarrow \overset{2}{\underset{1}{\bigcirc}} \xrightarrow{+\text{H}^+} \text{Pt}-\!\!\underset{1}{\overset{2}{\bigcirc}}$$

The H—D exchange in naphthalene occurs selectively in the β-position, which is accounted for by the intermediate formation of β-naphthyl derivatives.

The α-naphthyl platinum derivatives are apparently sterically hindered, retarding the 2 → 1 shift. It is less understandable, however, why the 2 → 3 shift does not occur: it is unlikely to be due merely to steric hindrance. Probably, the intermediate state of the type

is thermodynamically unfavourable.

Thus, the data on multiple exchange indicate the intermediate formation of complexes with carbenes and metallocycles together with alkyl platinum complexes

$$(V.7)$$

An accurate calculation of the relative rates of the elementary reactions from the data on multiple exchange for hydrocarbon containing more than two carbon atoms is difficult. However, the analysis of the distribution in the absence of steric hindrance shows that the rate constants decrease in the order $k_{11} \gg k_{12} > k_{13} > k_{14}$.

An analysis of the ascending distribution in the case of methane H–D exchange in aqueous solution from the viewpoint of several possible mechanisms shows that the reaction proceeds with the participation of the methylene complex, but no evidence for the formation of the Pt≡CH methine complex was obtained.

Another mechanism for multiple exchange has been proposed by Rudakov et al. [22]. The authors found that the multiple exchange factor M for H–D exchange of cyclohexane increases from 1 to 1.8 with an increase of acidity. Proceeding from the assumption that the dissociation rate

$$Pt{-}CH{<} \longrightarrow Pt{\cdots}C{<} + H^{+}$$

does not depend on H^+ concentration, the authors have come to the conclusion that the multiple exchange is due to the formation of the complex of Pt(II) with an alkane

$$Pt-CH{\Big<} + H^+ \rightleftharpoons [PtCH_2{\Big<}]^+$$

According to this hypothesis the rate of the complex decomposition to form free alkane must be comparable with its dissociation rate into the alkyl platinum complex and a proton

$$Pt(II) + RH \underset{k_{-1}}{\overset{k_1}{\rightleftharpoons}} Pt(II)RH \overset{k_2}{\longrightarrow} Pt-R + H^+$$

In steady-state conditions:

$$k_2[Pt(II)RH] + k_{-1}[Pt(II)RH] = k_1[Pt(II)][RH]$$

and, since both terms of the equation left side are comparable, the rate of the complex formation ($k_1[Pt(II)][RH]$) must be very slow and close to the total rate of H–D exchange. Moreover, the rate of the complex decomposition must be also rather slow, implying that it is comparatively stable. This seems unlikely for the reaction of alkanes with platinum(II) complexes.

It looks more plausible that the rate of the carbene–platinum complex formation increases with acidity under conditions of the work [22], e.g., as a result of protonation

$$\underset{}{HO}{\Big>}Pt{\Big<}^R + H^+ \rightleftharpoons \left[\underset{}{H_2O}{\Big>}Pt{\Big<}^R\right]^+$$

An increase of positive (or a decrease of negative) charge of the complex could increase the rate of the carbene formation in the reaction of the C–H bond dissociation in the Pt(II) coordination sphere to form H^+ and carbene.

V.2.5. OXIDATION AND DEHYDROGENATION OF ALKANES IN THE PRESENCE OF Pt(II) AND Pt(IV) COMPLEXES: PRODUCTS

In 1968 Hodges and Garnett [4], in a study of H–D exchange in benzene in the presence of platinum(II) complexes as catalysts, showed that, when H_2PtCl_6 is present, a homogeneous reduction of Pt(IV) by benzene takes place, with the formation of platinum(II), chlorobenzene, and small amounts of diphenyl:

$$H_2PtCl_6 + C_6H_6 \xrightarrow{Pt(II)} C_6H_5Cl + HCl + H_2PtCl_4$$

$$H_2PtCl_6 + 2\,C_6H_6 \xrightarrow{Pt(II)} C_6H_5C_6H_5 + 2\,HCl + H_2PtCl_4$$

The study of the H–D exchange of alkanes catalyzed by Pt(II) showed that when Pt(IV) was added it also oxidized acetic acid present as a solvent to chloroacetic acid [23, 24].

Based on the fact that the alkanes turned out to differ only quantitatively from benzene and acetic acid in H–D exchange with D_2O catalyzed by Pt(II) complexes, it could be concluded that alkanes too were capable of catalytically reducing H_2PtCl_6 in the presence of Pt(II) complexes. In fact, it was shown in 1972 [24] that alkanes did reduce Pt(IV) to Pt(II) under conditions similar to those of H–D exchange with the solvent (at $100-120°C$), Pt(II) being the necessary catalyst for the reaction. Assuming the formation of alkyl Pt complexes in the mechanism of H–D exchange of alkanes, their participation in reactions different from H–D exchange, taking into consideration the well-known reactivity of alkyl metal derivatives, was not altogether surprising [14, 23–25].

Table V.6 lists the products of the oxidation of alkanes obtained in the aqueous solution of $H_2PtCl_6-Na_2PtCl_4$ [23]. They are mainly chloroalkanes together with alcohols, ethers, ketones, and acids for linear hydrocarbons. Cyclic

TABLE V.6

Products of alkane oxidation in aqueous solution of $H_2PtCl_6-Na_2PtCl_4$

Nos.	Alkane	t °C	Time h	Products and ratios of their yields, %
1.	Methane (50 atm.)	120	4	Methyl chloride, methanol*
2.	Ethane (20 atm.)	120	4	Ethyl chloride, ethanol*
3.	Propane (6 atm.)	120	4	n-C_3H_7Cl (75), i-C_3H_7Cl (25)
4.	Pentane	120	0.25	n-$C_5H_{11}Cl$ (56), sec-$C_5H_{11}Cl$ (44)
5.	Isopentane	120	0.25	n-$C_5H_{11}Cl$ (78), sec-$C_5H_{11}Cl$ (22)
6.	Hexane	110	5	n-$C_6H_{13}Cl$ (40), n-$C_6H_{13}OCOCF_3$ (30.5), $C_4H_9COCH_3$ (23.9), sec-$C_6H_{13}OCOCF_3$, sec-$C_6H_{13}Cl$ (1.4)
7.	Cyclohexane	110	5	C_6H_6 (77.6), C_6H_5Cl (11), $C_6H_{11}Cl$ (5.4), $C_6H_{11}OCOCF_3$ (5), cyclohexanone, alcohol and traces of dichloride.
8.	Methylcyclohexane	110	5	Toluene (61.6), p-Cl-toluene (11.2), $CH_3C_6H_{10}Cl$ (9.1), $CH_3C_6H_{10}OCOCF_3$ (8.2), $C_6H_{11}CH_2Cl$ (4.1), ketone (4.1)
9.	Decalin	110	5	Naphthalene

Experiments 1–5 in H_2O, 0.5 M Pt(IV), 0.05 M Pt(II).

Experiments 6–9 in 8.7 M CF_3COOH, 0.18 M Pt(IV), 0.02 M Pt(II).

* Alcohols were not reported in the work [23] but later were found to be the products at least in some cases (see Sections V.2.6 and V.2.7).

alkanes (cyclohexane, decalin) produce high yields of aromatic hydrocarbons (benzene, naphthalene).

The dehydrogenation products (e.g., benzene from cyclohexane) can be accounted for by a primary dehydrochlorination. The olefins formed primarily are more reactive than the alkanes (evidently, because of the weakened C—H bond in an allyl position to the double bond) and are subjected to further chlorination-dehydrochlorination reactions. In the case of cyclohexane, the final product is stable benzene. In the cases of alkanes such as hexane, octane, and decane, the chloroalkane yield at 120°C is very low, whereas a considerable amount of Pt(IV) is reduced to Pt(II). For example, when oxidizing one hexane molecule, 5—6 molecules of the Pt(IV) complex are reduced to Pt(II), indicating that the primary products of the hydrocarbon reaction rapidly enter into subsequent reactions (probably those of chlorination-dehydrochlorination). The final products have not been identified [26].

It has been reported that in the presence of an external oxidant (air together with $CuCl_2$ or quinones) the $H_2PtCl_6 - PtCl_4^{2-}$ system can be made catalytic with respect to both Pt(IV) and Pt(II). However, its catalytic activity is still low. Thus, upon the controlled introduction of HCl in the reaction with acetic acid, chloroacetic acid was produced in amounts corresponding to five cycles for Pt [27].

Oxidation of methane in water solution by $H_2PtCl_6 - PtCl_4^{2-}$ produces CH_3OH and CH_3Cl. In the presence of heteropolyacid, air oxidizes methane catalytically but methanol itself is oxidized in this system, while gaseous CH_3Cl is evolved from the solution and can be accumulated [28].

In [29], a catalytic system involving K_2PtCl_4, H_2PtCl_6, and $HgSO_4$ (the latter is used as a Cl^- acceptor) deposited on silica was studied for the chlorination of alkanes, methane in particular. It is reported that, at 100°C, this system is 10^5 times more active with respect to the oxidative methane chlorination to produce CH_3Cl than the well-known copper chloride catalyst. On more extensive conversions, CH_2Cl_2, $CHCl_3$, and CCl_4 are also formed.

V.2.6. REACTIVITY OF DIFFERENT HYDROCARBONS. MECHANISM OF OXIDATION

A study of the reactivity of hydrocarbons of various structures with different C—H bonds demonstrates a considerable similarity between the reactions of oxidation and H—D exchange with the solvent.

It can be seen from Table V.6 for propane that the ratio between n-propyl chloride and isopropyl chloride is 3 : 1, which coincides with the ratio of the rates for H—D exchange of the primary and secondary C—H bonds. However, the ratio of isomers in the products might not always correspond to the ratio of rates of the chlorination due to the different rates of the secondary reactions, for example, hydrolysis. Hence, valuable information can be derived from the comparison of reaction rates of various hydrocarbons, measured by the decrease of their concentration due to oxidation (Table V.7) [30].

First of all it is essential to note the important role of steric factors which manifest themselves in a very similar way to the case of H–D exchange. Thus methyl and methylene groups adjacent to the quaternary carbon atom practically

TABLE V.7

Rate constants of the oxidation of hydrocarbons aqueous solutions of chloride complexes of platinum. $[K_2PtCl_4] = 5 \times 10^{-2}$ M; $[H_2PtCl_6] = 2 \times 10^{-2}$ M; $98°C$; k_{ox} and k_{exch} in s^{-1}, n_0 = the number of C atoms accessible to attack

Hydrocarbon	$k_{ox} \times 10^5$	n_0	$k_{ox} \times 10^5/n_0$	$k_{exch} \times 10^{5a}$
Methane	1.6	1	1.6	0.24
Ethane	6.6	2(2.0)	3.3	1.1
Propane	9.9	3(3.0)	3.3	2.0
n-Butane	11.5	4(3.5)	2.9	2.9
n-Pentane	15.3	5(4.6)	3.1	3.3
n-Hexane	15.9	6(4.8)	2.6	3.8
n-Heptane	16.5	7(5.0)	2.4	
n-Octane	15.4	8(4.7)	1.9	
2-Methylpropane	8.2	3	2.7	1.0
2,2-Dimethylpropane	(1.0)[b]	0	–	0.1
2-Methylbutane	11.2	4	2.8	1.8
2,2-Dimethylbutane		1		1.0
2,3-Dimethylbutane		2		1.2
2-Methylpentane		5		3.0
3-Methylpentane	12.0	5	2.4	2.7
2,2,4-Trimethylpentane	3.6	2	1.8	
Cyclopentane	21.1	5	4.2	5.6
Cyclohexane	30.0	6	5.0	7.6
Cyclohexane-d_{12}	20.3	6	3.4	
Methylcyclopentane	20.0	5	4.0	
Methylcyclohexane	18.8	6	3.1	
Benzene	35.8	6	6.0	(17.2)[c]
Toluene	31.5	6	5.3	
Ethylbenzene	40.0	7	5.7	
Isopropylbenzene	43.2	7	6.2	

[a] The values of k_{exch} correspond to 50 mole % in AcOD in D_2O, $[K_2PtCl_4] = 2 \times 10^{-2}$ M, $100°C$.
[b] An estimation based on correlation of k_{ox} and k_{exch}.
[c] Extrapolation from $80°C$.

do not participate in the oxidation. 2,2-Dimethylpropane and 2,2,3,3-tetra-methylbutane were shown to be oxidized by Pt(IV) much more slowly than alkanes with the same number of carbon atoms but with no quaternary carbon. According to [30], the number of carbon atoms (n_0) accessible to attack by the Pt compounds can be approximately determined by the formula

$$n_0 = n - n_{tert} - 5n_{quat}$$

where n is the total number of carbon atoms, and n_{tert} and n_{quat} are the numbers of tertiary and quaternary carbon atoms in the molecule, respectively. It is interesting to note that for normal alkanes, as seen from Table V.7, the value k/n decreases with increasing n. In the authors' opinion, this effect reflects the growth of steric barriers as a result of the hydrophobic molecule rolling itself up into a 'coil'. Table V.7 gives, in parentheses, the effective number of carbon atoms accessible to attack, determined as $k/3.3$ (the coefficient 3.3 was selected from the data for ethane and propane, where it corresponds to the number of carbon atoms in these molecules). For n-octane almost half the carbon atoms do not participate in the reaction. It should be noted that the rate constant of the oxidation remains practically unchanged, despite the increase of the number of carbon atoms from pentane to octane. This may mean that only the C—H bonds of the primary and secondary C atoms from hydrocarbon are reactive (i.e., ω- and (ω −1)-carbon atoms). If this is so, then steric factors in the alkane oxidation by Pt complexes produce the same effect as they do in the selective oxidation under the action of monooxygenase with the participation of cytochrome P-450 and methane monooxygenase. Only the primary CH_3 groups (if not directly attached to the quaternary carbon atom) and the secondary CH_2 group adjacent to the CH_3 group are active in the oxidation. The remaining methylene groups, bonded to longer alkyl groups other than CH_3, are practically inactive.

There is comparatively little difference between the rate constants relating to one C—H bond in the reactive CH_3 and CH_2 groups of different alkane molecules. Only the C—H bond in methane is somewhat less active that that in other hydrocarbons but, in this case also, the difference is only a factor of two. A comparison of the rate constant corresponding to different platinum species is presented in Table V.8 [31].

Similarly to H—D exchange, neither $[Pt(H_2O)]_4^{2+}$ nor $[PtCl_4]^{2-}$ is active in the oxidation of alkanes. The values of k_1, k_2, k_3 are not very different from each other. It is interesting that the positive doubly-charged ion shows an adequate activity for benzene, whereas the maximum activity is found for the singly-charged $[PtCl(H_2O)_3]^+$ ion. This evidently reflects the more classical electrophilic nature of the reaction displayed in the case of positively charged

TABLE V.8
Rate constants k_1 (M^{-1} s^{-1}) of the reaction of Pt^{2+} (k_0), $PtCl^+$ (k_1), $PtCl_2$ (k_2), $PtCl_3^-$ (k_3) and $PtCl_4^{2-}$ (k_4)

Hydrocarbon	k_0	k_1	k_2	k_3	k_4
Isobutane	0	1.9 ± 0.1	3.8 ± 1.1	3.7 ± 0.1	0
Cyclohexane	0	5.3 ± 0.2	9.5 ± 2.4	12 ± 0.3	0
Benzene	9.1 ± 1.4	66 ± 31	12 ± 4	18 ± 2	0

Pt complexes interacting with aromatic compounds. The stability of the addition products of Pt^{2+} and $ClPt^+$ to the aromatic ring makes it possible to carry out the reaction in two stages:

In the case of alkanes, the Pt–C bond formation and C–H bond cleavage should be synchronous processes (i.e. the 'electrophilic' addition of Pt to the carbon atom should be accompanied by 'nucleophilic' addition of Pt or one of its ligands to the proton). Hence, the positive charge is not of particular importance here. It is also interesting to note that according to the data of Table V.8, the reaction rate constants of the less electrophilic species – the uncharged platinum complex $PtCl_2(H_2O)$ and the negative ion $[PtCl_3H_2O]^-$ – are practically identical for benzene and cyclohexane and that the ratio of the rate constants of the reactions of cyclohexane and isobutane remains constant for the reaction with different platinum complexes present in solution.

A large number of similarities existing in the reactions of alkane H–D exchange with the solvent and of alkane oxidation under the action of Pt(IV) (e.g., the similarity in the selectivities of attack in reactions of various C–H bonds) leads us to the conclusion that the same stage is involved in both cases. This stage must include the alkane's reaction with the Pt(II) complex acting as a catalyst. It is natural to assume that the common stage must be the formation of alkyl platinum derivatives, which are intermediates in the oxidation and exchange reactions.

The kinetic data confirm this assumption. A comparison of the rate constants of H–D exchange and hydrocarbon oxidation reveals the direct correlation between them [30]. Under certain conditions the rate constants of oxidation and exchange and their activation energies may be practically the same (see Table V.9).

TABLE V.9

Comparison of rate constants and activation energies for exchange and oxidation
in $CH_3COOD-D_2O$

Hydrocarbon	$k \times 10^4$ (M^{-1} s^{-1}), 100°C		E_0 (kcal mole^{-1})	
	exchange	oxidation	exchange	oxidation
Benzene	60	55	25.7	25
Acetic acid	0.9	1.25	23.7	22.5
Hexane	18	16	–	–

Direct evidence for a common stage in the oxidation and exchange of alkanes
in the presence of Pt(II) complexes has been obtained [32], where both the
reactions were studied jointly at different concentrations of acid and Pt(IV)
complexes (see Table V.10). The total of the exchange and oxidation rates was
shown to remain constant and independent of the concentration of acid and
Pt(IV), but their ratio changes with a change of its concentration; with increasing Pt(IV) concentration and constant acid concentration, the rate of
the oxidation process increases, whereas, with increasing acid concentration
(and constant Pt(IV) concentration), there is an increase in the rate of the

TABLE V.10

Effect of the oxidant and trifluoroacetic acid on k_{ox} and k_{exch} of cyclohexane in
solutions of chloride complexes of platinum. $[K_2PtCl_4] = 5 \times 10^{-2}$ M, 98°C

$[CF_3CCOD]$ or $[CF_3COOH]$ M	$[D_2PtCl_6]$ or $[H_2PtCl_6] \times 10^3$ M	$k_{ox} \times 10^5$ s^{-1}	$k_{exch} \times 10^5$ s^{-1}	$(k_{ox} + k_{exch}) \times 10^5$ s^{-1}
0	0	24.2	2.5	26.7
0	20	28.0	0.2	28.2
0	20	28.4	–	28.4
0	20	28.1	–	28.1
2	0	7.2	12.0	19.2
2	1	5.4	11.3	16.8
2	5	13.0	3.1	16.1
2	10	15.3	1.3	16.6
2	20	17.7	0.65	18.3
2	20	18.7	–	18.7
4	0	0.8	12.6	13.4
4	20	9.0	2.2	11.2

exchange reaction. These data are in agreement with the common mechanism of formation of alkyl Pt derivatives in the first stage:

$$Cl_2Pt^{II} + RH \longrightarrow RPtCl + H^+ + Cl^-$$

followed by two competing reactions

$$RPtCl \left\{ \begin{array}{l} \xrightarrow{\;H^+,\,Cl^-\;} RH + PtCl_2 \\ \xrightarrow{\;Pt(IV)\;} \text{oxidation} \end{array} \right.$$

The kinetics of the catalytic chlorination of the C—H bond have been studied extensively for the case of the $PtCl_6^{2-}$ reaction of acetic acid [33]. Similar to the hydrocarbon reactions, the reaction is catalyzed by Pt(II) chloride complexes and proceeds almost quantitatively according to the scheme:

$$PtCl_6^{2-} + CH_3COOH \xrightarrow{\;PtCl_4^{2-}\;} PtCl_4^{2-} + ClCH_2COOH + HCl$$

In the absence of Pt(II), the Pt(IV) concentration decreases autocatalytically, the observed induction period being removed by the addition of bivalent platinum. The reaction is convenient for kinetic study (because of the absence of by-products) and obviously proceeds according to a mechanism similar to that for the oxidation of alkanes. The reaction is first order with respect to the Pt(II) and acetic acid concentrations, and is retarded on the addition of acid and Cl^- ions, its rate being inversely proportional to the square of chloride-ion concentration at high Cl^- concentrations. The order of reaction with respect to Pt(IV) changes from 0 to 1. The mechanism suggested for this reaction is evidently common for various C—H bond containing compounds including saturated hydrocarbons

$$PtCl_4^{2-} \rightleftharpoons PtCl_3^- + Cl^- \tag{V.8}$$

$$PtCl_3^- \overset{k}{\rightleftharpoons} PtCl_2 + Cl^- \tag{V.9}$$

$$PtCl_2 + RH \underset{k_{-1}}{\overset{k_1}{\rightleftharpoons}} RPtCl + H^+ + Cl^- \tag{V.10}$$

$$RPtCl + PtCl_6^{2-} \xrightarrow{k_2} RCl + 2\,PtCl_3^- \tag{V.11}$$

The rate-determining steps may be Reactions V.10 or V.11, depending on the conditions, since Reactions V.8 and V.9 proceed rapidly to reach equilibrium. At temperatures below 100°C and at sufficiently high platinum concentrations, the reaction rate depends neither on Pt(IV) concentration nor on acidity and is apparently determined by reaction V.10 of the substrate interaction with Pt(II). At higher temperatures and low Pt(IV) concentrations, there appears a dependence on Pt(IV) concentration (up to direct proportionality) and on the acidity (inverse proportionality). In this case the rate determining step is stage V.11, the Pt(II) alkyl derivative interaction with Pt(IV).

The following kinetic equation based on the scheme above can be obtained using the steady-state concentrations method and assuming reaction V.8 shifted to the right:

$$v = \frac{k_1 k_2 K [RH]\ [Pt^{II}]\ [Pt^{IV}]}{([Cl^-] + K)(k_{-1} [H^+]\ [Cl^-] + k_2 [Pt^{IV}])} .$$

The equation is in good agreement with the experimental data. The temperature dependence of the reaction rate corresponds to two activation energies. Over the temperature range of 80–100°C the effective activation energy E_{eff} = 22.5 kcal mole^{-1}, and over the range of 100–120°C, E_{eff} = 8.6 kcal mole^{-1}. At a temperature below 100°C and low chloride concentration, when the reaction rate is practically independent of [Cl$^-$], the equilibrium (V.9) seems to be almost completely shifted to the right and we can disregard [Cl$^-$] in comparison with K and k_1 [H$^+$] [Pt^{II}] as compared with k_2 [Pt^{IV}]. In this case, the kinetic equation corresponds to

$$v = k_1\ [RH]\ [Pt^{II}]$$

In this temperature range the activation energy of the oxidation is close to that of the exchange.

At temperatures above 100°C, the reaction of Pt(II) alkyl derivatives with Pt(IV) becomes the rate-determining step. The low values of both the activation energy and the preexponential factor correspond to the nature of process (V.11), which is evidently complicated and involves, as a first stage, the formation of a Pt(IV) alkyl derivative with a subsequent reductive elimination.

$$\longrightarrow RPtCl_3 + PtCl_4^{2-}$$

$$RPtCl_3 \longrightarrow RCl + PtCl_2$$

A transfer of Cl atoms from Pt(IV) to Pt(II) or a transfer of an alkyl group from Pt(II) to Pt(IV) takes place in a binuclear activated (or intermediate) complex. This mechanism is in agreement with the higher oxidation ability of benzene [14] and alkanes [34] in systems involving Br$^-$, which forms bridging bonds more readily than does Cl$^-$.

V.2.7. THE ISOLATION OF ARYL AND ALKYL PLATINUM COMPLEXES IN
REACTIONS OF PLATINUM HALIDES WITH HYDROCARBONS AND THEIR
ROLE AS INTERMEDIATES

The experimental data reported in the previous section leave no doubt of the intermediate formation of aryl and alkyl platinum derivatives in the Pt(II) reaction with arenes and alkanes. Moreover, some of the data show the high energy of the Pt—C bond to be the driving force of the reaction. For example, the comparatively small differences in the reaction rate constants of various alkanes, including methane and its homologues, may be explained by the fact that differences in the C—H bond energies are compensated by corresponding differences in energies of Pt—C bonds formed. Nevertheless, the reaction conditions (comparatively high temperature, protic media), and the high reactivity of metal alkyl derivatives could not inspire much hope for the possibility of observing intermediate alkyl and aryl platinum complexes, and even less so, for their isolation from the solution. However, the reaction provided one more unexpected result for its investigators. Intermediate aryl and some of the alkyl derivatives turned out to be sufficiently stable to be formed in considerable concentrations in the reaction mixture. It was possible to detect them by NMR and, in some cases, even to isolate them from the solution in crystalline form.

The first results which could be interpreted in terms of the formation of appreciable concentrations of aryl and alkyl Pt(IV) derivatives were reported in [14, 26, 27], where the kinetics of H_2PtCl_6 consumption in the hydrocarbon oxidation together with the accumulation of reaction products was studied. If the reaction is carried out in the presence of Pt(II), the consumption of H_2PtCl_6 starts immediately, with the maximum rate being achieved already at the initial stage. At the same time the main reaction products, such as chlorobenzene and diphenyl in benzene oxidation, are accumulated throughout the induction period at a considerably lower initial rate.

Meanwhile, there appears to be an intermediate in the solution which is a Pt(IV) complex which forms benzene when heated with a reducing agent ($SnCl_2$, $NaBH_4$, H_2 or $N_2H_4 \cdot H_2O$). The concentration of this complex (determined by the benzene formation in the reaction with hydrazine hydrate) initially rises and then falls.

Similar results have been obtained with acetic acid [27]. The authors of [26, 27] came to the conclusion that the observed organic derivatives of Pt(IV) are compounds with a Pt—C σ-bond (Pt—C_6H_5 and Pt—CH_2COOH, respectively). This conclusion, which postulates a striking stability of the Pt—C bond, was confirmed by NMR [35]. The NMR spectrum of the solution, in which the benzene reaction with H_2PtCl_6 was carried out in the presence of Pt(II), showed the formation of a compound with a Pt—Ph bond.

The isolation of aryl complexes of platinum(IV)

Later, in studies specially set up to investigate the reaction of H_2PtCl_6 with a number of aromatic compound [36–38], it was possible to isolate these aryl derivatives in crystalline form, and at least for two cases (naphthalene and *o*-nitrotoluene) to determine their structure by X-ray analysis (Figures V.5 and V.6).

Figure V.5 shows that complexes formed from naphthalene are β-naphthyl derivatives with normal Pt—C bonds. This is in good agreement with the data on naphthalene exchange in the presence of Pt(II) complexes obtained earlier (see above), which have indicated that the H—D exchange occurs on the β-carbon atom of naphthalene. The formation of the α-naphthyl derivative is apparently sterically hindered. The complexes isolated are in the form of ammonium salts and contain ammonia in a *trans*-position to the aryl group. It was shown that ammonia was present in the adsorbent of the chromatographic column, and apparently ammonia replaces a water molecule in the Pt(IV) coordination sphere, whereas NH_4^+ replaces the counter ion, i.e., evidently the complexes produced in the reaction are compounds of the formula $(H_3\overset{+}{O})\,[ArPtCl_4(H_2O)]^-$.

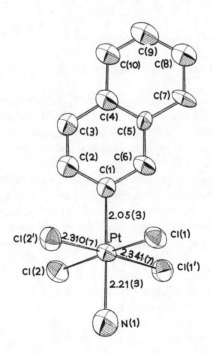

Fig. V.5: The structure of the Pt(IV) complex produced in the reaction with naphthalene.

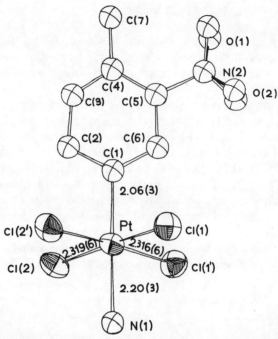

Fig. V.6: The structure of the Pt(IV) complex produced in the reaction with *o*-nitrotoluene.

Thus, the reaction of metallation using H_2PtCl_6 turned out to be the most convenient method for the synthesis of the Pt(IV) aryl derivatives which have now become accessible for various aromatic molecules. Besides, a study on the formation and properties of aryl complexes throws additional light on the reaction with hydrocarbons.

The complex formation is catalyzed by additions of Pt(II) compounds, which react with arenes producing rather unstable intermediate Pt(II)-aryl complexes. The latter, however, have also been isolated in the crystalline state for the cases of benzene and toluene. The aryl Pt(II) complexes have been shown to form Pt(IV) aryl complexes on reacting with H_2PtCl_6. A further reaction of aryl Pt(IV) complexes produces the same products as in the oxidation of the corresponding arenes. Thus, the *p*-anisyl Pt(IV) complex gives chloroanisole and dianisyl, i.e. the same products which are formed in the anisole oxidation. Thus each step of the sequence of reactions

$$Ar\!-\!H + Pt(II) \longrightarrow Ar\!-\!Pt(II) \xrightarrow{\ H_2PtCl_6\ } Ar\!-\!Pt(IV) \longrightarrow ArCl,\ Ar\!-\!Ar$$

suggested from earlier studies, was directly demonstrated. The formation of aryl platinum(IV) complexes is determined to a considerable degree by steric factors (a situation analogous to that observed in H–D exchange and oxidation). No o-derivatives are formed in the interaction with substituted benzene, hence some di- and polysubstituted compounds are practically not metallated by Pt(IV) in the aromatic ring (e.g., no aryl Pt(IV) derivatives were observed in the cases of p-xylene, mesitylene, pentafluorobenzene, and pentamethylbenzene).

Arenes with electropositive substituents react more readily than benzene, while electronegative ones slow down the reaction. The reactivity of different aromatic compounds tested in the reaction with H_2PtCl_6 changes in the following order: phenol > anisole > naphthalene > toluene > ethylbenzene > diphenyloxide > isopropylbenzene > diphenyl > benzene > fluorobenzene > acetophenone > benzoic acid > chlorobenzene > nitrobenzene. In the case of toluene, a p-tolyl derivative of Pt(IV) is initially produced (about 90%) (i.e. the reaction follows the rules of electrophilic substitution). However, the subsequently produced complex isomerizes relatively rapidly with the formation of products with a statistical distribution of p- and m-substituents. It may be noted that, in Pt(II)-catalyzed chlorobenzene chlorination using H_2PtCl_6, a statistical mixture of m- and p-isomers is produced.

Intermediate methyl platinum complex in the reaction of H_2PtCl_6 with methane [39, 40]
Success in the isolation of aryl complexes of Pt(IV) from arenes has allowed an attempt to isolate an intermediate complex in the reaction with methane, the most inert hydrocarbon among alkanes. The study of methane oxidation at $120°C$ under the action of H_2PtCl_6 in the mixture of water and trifluoroacetic acid (CH_4 pressure *ca.* 100 atm.) showed the solution to contain a platinum complex producing methane if decomposed in the presence of a reducing agent, such as, for example, hydrazine hydrate or sodium borohydride.

When decomposing this complex by DCl in D_2O, CH_3D is produced. Upon heating in water the complex produces methyl chloride and methanol, i.e. the same products as those formed in methane oxidation. Using kinetic data it was possible to select optimum conditions to get sufficient concentrations of the intermediate complex, which was isolated by the addition of triphenylphosphine as neutral $CH_3Pt(Ph_3P)_2Cl_3$ [39].

The kinetics of methyl platinum(IV) complex concentration change were followed in [40], together with the decrease of Pt(IV) concentration and the increase of the products' (methyl chloride and methanol) concentration in the reaction of $PtCl_6^{2-}$ with methane in water (Fig. V.7). The reaction rate

$$-\frac{d[Pt(IV)]}{dt} = \frac{d([CH_3Cl] + [CH_3OH])}{dt}$$

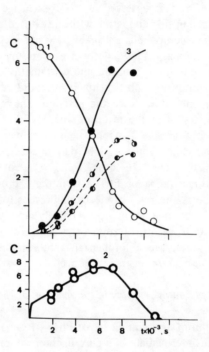

Fig. V.7: The kinetics of $PtCl_6^{2-}$ oxidation of methane in water solution in the presence of $PtCl_4^{2-}$, $t = 120°C$, $P_{CH_4} = 100$ atm, $[PtCl_4^{2-}]_0 = 1.5 \times 10^{-2}$ M; curve 1 – $[PtCl_6^{2-}] \times 10^2$ M; curve 2 – $[CH_3PtCl_5^{2-}] \times 10^5$ M; curve 3 – $[CH_3OH + CH_3Cl] \times 10^2$ M; ◑ – $[CH_3OH] \times 10^2$ M; ◐ – $[CH_3Cl] \times 10^2$ M.

was found to be proportional to the methyl platinum(IV) complex concentration, in agreement with the suggestion that the latter is the intermediate. But the most direct proof that the Pt(IV)—CH_3 complex is the sole intermediate of the reaction came out of measurements of its decomposition (which produces CH_3Cl and CH_3OH, i.e. the same products as in methane oxidation). The rate of the decomposition measured at different temperatures when extrapolated to the temperatures of the reaction of $PtCl_6^{2-}$ with CH_4 catalyzed by Pt^{II} exactly coincides with the total reaction rate for the concentration of Pt^{IV}—CH_3 observed in the reaction mixture. Thus, knowing the decomposition rate constant measured separately and the concentration of Pt(IV)—CH_3 complex, the reaction rate of methane oxidation by $PtCl_6^{2-}$ can be correctly predicted, leaving no doubt that the Pt(IV)—CH_3 complex is the sole intermediate.

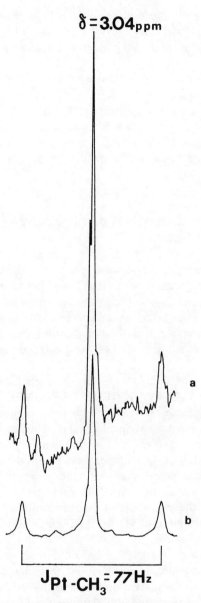

Fig. V.8: PMR spectra of the MePt^{IV} complex obtained in (a) the PtCl$_6^{2-}$ (0.32 M) + PtCl$_4^{2-}$ (0.24 M) + CH$_4$ (0.1 M) system, time of reaction 30 min. at 120°C, number of scans 4 × 10^3; (b) the reaction of MeI + PtCl$_4^{2-}$, complex concentration [Pt(IV)−CH$_3$] = 2 × 10^{-2} M, number of scans 1 × 10^2, solvent − D$_2$O.

The complex apparently identical to this can be obtained in high concentrations. when reacting CH_3I with $PtCl_4^{2-}$ [41]. Its PMR spectrum is presented in Fig. V.8 with the line at 3.04 p.p.m. corresponding to $Pt(IV)-CH_3$ protons with the satellites due to the presence of ^{195}Pt isotope (33.8% in natural Pt). The rate constant of the complex decomposition in water solution at 60–100°C turned out to be equal to that of the methylplatinum(IV) intermediate in methane oxidation. The PMR spectrum of the latter was shown [40] to contain the same lines (see Fig. V.8) as for the complex prepared from CH_3I and $PtCl_4^{2-}$. Thus, because of the striking stability of methylplatinum(IV) complex, its intermediate formation in methane oxidation was established with full certainty.

V.2.8. GENERAL CONSIDERATIONS OF THE REACTIONS OF HYDROCARBONS WITH Pt(II) COMPLEXES

The interaction with Pt(II) complexes was the first example of alkane reactions with transition metal complexes in which formation of the metal-carbon bond was unambiguously established. Hence this reaction deserves special consideration. Although the Pt(IV)-alkyl bond is more stable than Pt(II)-alkyl one (at least, in the kinetic sense), it is the coordinatively unsaturated derivatives of platinum(II) which react with alkanes, octahedral complexes of Pt(IV) being inactive.

Among the different Pt(II) complexes, those containing ligands which are hard bases, such as H_2O, Cl^-, CH_3COO^-, CF_3COO^-, are the most reactive towards alkanes. Evidently, such complexes form comparatively strong bonds of Pt^{2+} with such a soft base as an alkyl anion. The introduction of other soft bases as ligands (phosphines, CN^-, I^-, etc.) apparently draws off the platinum orbitals in the direction of these ligands and weakens the Pt—C bond. Thus, the strength of Pt—C (and probably Pt—H) bonds is apparently the main driving force for the reactions of alkanes with Pt(II) complexes.

We may consider the mechanism of the reaction first from the point of view of electrophilic substitution of H^+ by Pt(II) complexes. This approach looks plausible in the case of arenes.

We can speak now of the 'platination' of aromatic compounds, which proceeds similarly to electrophilic mercuration, thallation, and plumbation (see Section I.1.3). The only important difference is that aryl derivatives of bivalent platinum are rather unstable; and aryl platinum complexes can be stabilized by converting them into derivatives of platinum(IV). The reaction of $Pt(H_2O)_4^{2+}$ with benzene can be suggested to proceed via the intermediate formation of a relatively stable cation in the initial stage which, in a later stage, reacts with the base B with proton elimination

Such a mechanism does not look attractive, however, for the case of alkanes, where the most electrophilic species, $Pt(H_2O)_4^{2+}$, is altogether unreactive. The reactive species, S_3PtCl^+, S_2PtCl_2 and, particularly, $SPtCl_3^-$, in aqueous solution, cannot be regarded as a strong electrophilic species similar to, e.g., superacids that react with alkanes to give carbonium ions.

What, then, is the mechanism of interaction of platinum(II) complexes with alkanes? It must undoubtedly include the synchronous C—H bond cleavage and formation of new Pt—C and Pt—H (or L—H) bonds. Only in this way can one explain the relative insignificance of polar effects and the importance of platinum-carbon bond energy as the driving force of the reaction.

Considering the experimental data obtained, particularly the formation of alkyl platinum species, the two most probable mechanisms of elementary interaction of Pt(II) with the alkanes may be suggested.

A. *Oxidative addition followed by proton abstraction* [2a] :

$$XPt^{II} + RH \longrightarrow XPt^{IV}\begin{matrix} R \\ \\ H \end{matrix} \longrightarrow Pt^{II}-R + HX$$

B. *Concerted substitution with the ligand participation* [42] :

$$XPt^{II} + RH \longrightarrow \begin{matrix} Pt------R \\ \vdots \qquad \vdots \\ X------H \end{matrix} \longrightarrow Pt^{II}-R + HX$$

For both mechanisms we can expect that there must be some optimum in denor and acceptor properties of the platinum complex. In both cases each reagent is simultaneously a donor and an acceptor of electrons and the net effect may be the comparative insignificance of polar factors. The other types of mechanisms proposed in the literature are less probable.

The mechanism of oxidative addition with the participation of a square planar Pt(II) complex is very attractive, since it has many analogies to the other reactions of complexes with a d^8 electronic configuration (such as oxidative addition of H_2). Direct observation of an oxidative addition of alkanes to Ir(I) (d^8) complexes was recently achieved (see Section V.3.3). The other analogue of the oxidative addition of RH to the Pt(II) complexes may be the reaction of the singlet methylene insertion into a C—H bond of alkanes, e.g.,

$$CH_2 + CH_4 \longrightarrow C_2H_6$$

considered in Section II.3. Calculations of the potential energy surface lead to the conclusion that the insertion proceeds in two stages of one elementary process: (1) approach of CH_2 to CH_4 along the methane C—H bond with a transfer of an H atom to CH_2; and (2) formation of the C—C bond proceeding similarly to the recombination of two methyl radicals.

As mentioned earlier (see Section I.2.5), there exists a number of examples of the positive interaction at the approaching of the low-valent metal ion to a C—H bond in a ligand ('soft hydrogen bonding'). Therefore, the approach of the alkane to the Pt(II) complex with the C—H bond moving along z axis might occur without a significant increase in the potential energy down to small values of platinum-carbon distances to start the platinum—carbon bond formation. The important difference between the Pt(II) complexes and CH_2 interactions with alkanes is that the Pt—H bond is apparently considerably weaker than the C—H bond, and, hence, the oxidative addition of RH to Pt(II) must be closer to a genuine insertion into the C—H bond, with the Pt—C bond starting to form earlier than a complete C—H bond cleavage.

The mechanism B, of replacement via the four-membered activated complex, has the advantage of direct formation of the Pt(II) alkyl complex without intermediate formation of $Pt^{IV}\diagdown{}^{R}_{H}$. It may be assumed that the alkyl hydride complex of Pt(IV) is a strong acid and the dissociation

$$Pt^{IV}\diagup\diagdown{}^{H}_{R} \longrightarrow [Pt^{II}-R] + H^+$$

is thermodynamically favourable in water solution, since it is also an oxidation of the hydride ion by tetravalent platinum. Possibly the real mechanism of an elementary process of alkane interaction with Pt(II) complexes represents a combination of processes A and B: the reaction may start as a Pt insertion in the C—H bond and terminate by the elimination of an HL species, e.g., H_2O or H_3O^+, without the intermediate formation of a stable alkyl hydride complex of Pt(IV).

Recently, the Pt(II) complex interaction with alkanes has been studied theoretically. Taking into account that Pt(II) behaves as an acceptor towards alkanes and the energy of the p_z free orbital may be too high, Shestakov [43] considered the symmetry effect of the ligand environment in platinum complexes on the mixing of a free orbital $(d_{x^2-y^2})$ and an occupied orbital (d_{z^2}).

It is postulated that, in order to react, the alkane molecule must approach the platinum complex along the z axis and the probability of the reaction increases with an increase of the admixture of d_{z^2} orbital to free orbital $d_{x^2-y^2}$. This leads to the conclusion that both the symmetrical platinum ions, those of $PtCl_4^{2-}$ and $[Pt(H_2O)_4]^{2+}$, must be inactive towards the alkanes, since there is no such admixture in both the cases and the maximum activity should appear in the case of $PtCl_2(H_2O)_2$, in the *trans*-form. As we have seen, these conclusions are in agreement with the experimental data.

The same conclusion was supported by a later study carried out using the extended Hückel method [44]. The authors suggested that the active state of the Pt(II) complex with respect to the alkanes was achieved by its transition to a non-planar state (Fig. V.9). With a decrease in the angle of deformation, Cl—Pt—Cl, the donor d_{xz} orbital moves upward while the acceptor $d_{x^2-y^2}$ orbital moves downward, since the antibonding interaction with ligands in the *OXZ* plane for $d_{x^2-y^2}$ is decreased on the distortion of plane-square geometry, and at the same time, it is increased for the d_{xz} orbital. Therefore, for angles $\varphi > 0$ the Pt(II)L_4 complex becomes both a better donor and a better acceptor. Here, the donor orbital of the complex correlates in symmetry with the antibonding orbital of the C—H bond (located in the *OXZ* plane), and the acceptor orbital of the complex correlates in symmetry with the bonding σ-orbital of the C—H bond.

Further consideration shows that, if we assume that the reaction of platinum with the alkane proceeds as a donor—acceptor interaction (with the electron density moving from the C—H bond to the Pt atom), then the reactivity of the *trans*-$PtCl_2(H_2O)_2$ must be highest among the Pt(II) complexes, since it is the best acceptor. The complexes $Pt(H_2O)_4^{2+}$ and $PtCl_4^{2-}$ must be considerably less active, in agreement with experimental evidence.

The hypothesis [44] looks attractive, in particular since it postulates the necessity of the platinum complex activation for the reaction of alkanes (by the distortion of its square planar configuration) which might correspond to the major contribution to the activation energy of the process. It could explain why, for a comparatively high activation energy (requiring the reaction to be carried out at elevated temperatures), the reactivity of various alkanes with considerable differences in the C—H bond dissociation energies differs rather insignificantly.

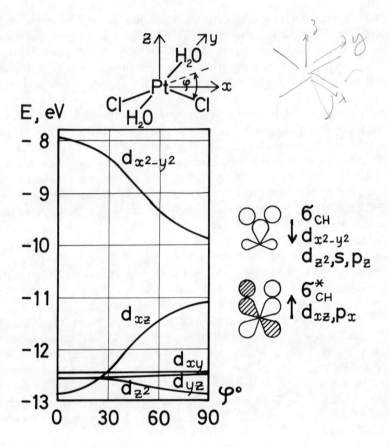

Fig. V.9: The change of orbitals energy in the complex Pt(II) on distortion of the Cl–Pt–Cl angle.

V.3. The Interaction of Alkanes with Complexes of Other Platinum-Group Metals

V.3.1. H–D EXCHANGE

Soon after the detection of the H–D exchange of alkanes with protic solvents with the participation of Pt(II) complexes, Garnett *et al.* [45] discovered that Ir(III) complexes are homogeneous catalysts of the H–D exchange of alkylbenzene with a solvent at 130°C.

Hydrogen atoms of both the benzene ring and the side chain are subject to exchange. Garnett reported also on the H–D exchange of alkanes with a solvent [46].

Na_3IrCl_6 was used as a catalyst, the solvent was a mixture of CH_3COOD and D_2O. Optimum results were obtained with a lower content of acetic acid than in the case of Pt(II) (the extent of cyclohexane exchange at 170°C within seven days increased from 0 to 54% when the acetic acid concentration was reduced from 38.5 mole % to 9.8 mole %).

The reaction proceeds more slowly than with Pt(II) complexes, but the complex is more stable toward heating and the reaction can proceed for a long time at high temperatures (150–170°C). Aromatic compounds (e.g., pyrene) are added to prevent metal participation.

A multiple exchange takes place ($M = 5.4$ for cyclohexane). Hydrogen atoms at a primary carbon atom undergo exchange most readily.

Rhodium trichloride has been shown [47] to catalyze the cyclohexane homogeneous exchange with a solvent (in the mixture of CH_3COOD–D_2O). An appreciable reaction rate is observed for a long time at a high temperature (96 hours, 130°C). The thermal stability of $RhCl_3$ (intermediate between those of Pt(II)) and Ir(III)) allows the reaction to proceed at high temperatures.

The results obtained for $RhCl_3$ are similar to those for Ir(III) and Pt(II): the exchange is multiple; benzene reacts more rapidly than cyclohexane; and for substituted benzenes, a steric effect prevents exchange in the *ortho*-position on the substituent.

The mechanism of H–D exchange in the presence of Ir(III) and Rh(III) complexes is probably close to that of Pt(II) catalysis. It may be that the real catalysts are complexes of Ir(I) and Rh(I) formed at high temperatures with the same electronic configuration (d^8) as the platinum complexes.

V.3.2. THE DEHYDROGENATION OF ALKANES BY Ir(I) COMPLEXES

An interesting reaction of cycloalkanes was recently discovered by Crabtree *et al.* [48] in solutions of cationic rhodium complexes in chlorohydrocarbons. $[IrH_2(acetone)_2L_2][BF_4]$ (L = PPh_3) may be a starting complex. This complex reacts readily with olefins to give $[Ir(olefin)_2L_2][BF_4]$ complexes. In excess cyclic olefin, the latter is hydrogenated, the iridium complex being converted into a species containing a coordinated diolefin. For example, the complex with cyclooctene after reflux in CH_2Cl_2 (40°C) for 30 min., changes into a complex with cyclooctadiene (75% yield), with formation of cyclooctane in solution. Thus, the hydrogen from the C–H olefin bond is used to hydrogenate one of the complexed olefin molecules. The driving force of the reaction is an entropy increase upon conversion of one of the olefin molecules to an alkane with its removal from

the complex into the solution. Thus, under relatively mild conditions, a number
of reactions, besides the one with cyclooctene described above, may take place.

$IrH_2(acetone)_2(PPh_3)_2$ +

$IrH_2(acetone)_2(PPh_3)_2$ +

$IrH_2(acetone)_2(PPh_3)_2$ +

It is interesting to note that in the formation of diene molecules from cyclooctene and bicyclooctene, the non-activated C—H bonds react. This implied the
possibility of involvement of cycloalkanes in an analogous reaction; which
turned out actually to be the case. Thus, cyclopentane in the same system forms
the cyclopentadienyl complex B (30% yield by PMR) after 18 hours in refluxing
1,2-dichloroethane containing 3,3-dimethyl-1-butene as a hydrogen acceptor.
Under similar conditions cyclooctane gives a cyclooctadiene complex with
70% yield over 4 hours. 3,3-Dimethylbutene is a necessary hydrogen acceptor
in this reaction as well. Other olefins inhibit the process with alkanes by forming
stable complexes with Ir(I). 3,3-Dimethylbutene apparently forms an unstable
complex and initiates a reaction with alkanes. The reaction appears to be entirely
homogeneous.

A study of iridium 1,5-cyclooctadiene complexes, such as $[Ir(cod)_2]^+$ and
$[Ircod(PR_3)_2]^+$, which are active homogeneous catalysts of olefin hydrogenation
[49] shows the metal ion to exhibit strong acceptor properties, which ensure its
catalytic activity. The oxidative addition of hydrogen halides to the complex proceeds in two stages, the first being the addition of a negative halide ion, which is
only then followed by the H^+ addition. If donor ligands (pyridine or Cl^-) are
introduced into the complex instead of acceptor phosphine ligands, the catalytic
activity disappears. Evidently, the acceptor properties of the iridium complex
in organic solution (e.g., dichloroethane) are considerably higher than those of
the platinum(II) complexes in aqueous solution. The acceptor nature of the
iridium complexes, which combine the properties of a Lewis acid and the
availability of d-electrons, seems to favour the C—H bond activation in alkanes.
The unstable solvate complex of IrL_2^+ with the solvent provides minor steric
hindrance in the reactions with alkanes, where the complex is likely to react in

a coordinatively unsaturated state. This is confirmed by the inertness which is apparently shown towards the iridium complex by solvents containing C—H bonds, such as methylene chloride or dichloroethane. This inertness is evidently due to the presence of an electronegative chlorine atom in the molecule. It has not yet been reported whether an activation of the C—H bond under the action of the iridium complex can be observed in linear hydrocarbons (it might be that the reaction proceeds effectively only with cycloalkanes).

V.3.3. OXIDATIVE ADDITION OF C—H BONDS IN ALKANES TO PHOTOCHEMICALLY GENERATED IRIDIUM(I) COMPLEXES

The first direct observation of oxidative addition of C—H bonds in alkanes to Ir(I) complexes was made recently in two studies [50, 51]. In both cases the active species are the result of the irradiation of initial Ir complexes in solution in a hydrocarbon (RH)

$$\underset{Me_3P}{\overset{Cp'}{>}}IrH_2 \quad \xrightarrow[+RH]{h\nu,\ -H_2} \quad \underset{Me_3P}{\overset{Cp'}{>}}Ir\underset{H}{\overset{R}{<}} \qquad [50]$$

$$Cp'Ir(CO)_2 \quad \xrightarrow[+RH]{h\nu,\ -CO} \quad \underset{OC}{\overset{Cp'}{>}}Ir\underset{H}{\overset{R}{<}} \qquad [51]$$

$$Cp' = \eta\text{-}C_5Me_5$$

The formation of hydroalkylmetal complexes was detected by their NMR spectra.

Presumably irradiation produces a 16-electron iridium(I) intermediate ($Cp'IrPMe_3$ or $Cp'IrCO$) to which the carbon—hydrogen bonds of alkanes add oxidatively. Alkanes with 'non-activated' C—H bonds such as neopentane and cyclohexane are active in both cases. Thus Ir(I) intermediates are apparently more active towards C—H bonds in alkanes than (η-C_5H_5)$_2$W obtained in a similar way from the dihydride complex, tungstenocene reacting only with 'activated' C—H bonds (see Section I.1.4).

The active complex ($Cp'IrPMe_3$) can be also produced in a thermal reaction. When the hydridoalkyl complex $Cp'(PMe_3)Ir(C_6H_{11})H$ is heated in benzene-d_6 or pentane at 110°C, the corresponding deuterophenyl deuteride or n-pentyl hydride are formed respectively [50].

The reactivity of Ir(I) active species to different types of C—H bonds, as shown in the work of Janowicz and Bergman [50], does not differ very much. Thus Cp'IrPMe$_3$ reactivities relative to cyclohexane are

Cyclohexane	1	Neopentane	1.14
Benzene	4	Cyclodecane	0.23
Cyclopropane	2.65	Cyclooctane	0.09

The *prim/sec* ratios for the reactions with propane and pentane are 1.51 and 2.7, respectively, while *tert*-C—H bonds are apparently completely unreactive. The isotope effect k_H/k_D for insertion of Cp' IrPMe$_3$ into the C—H (and C—D) bond in cyclohexane is as small as 1.38.

On the whole there is some similarity between the reactions of alkanes with Ir(I) and Pt(II) complexes, which supports the view that both have much in common as to the mechanism of their primary reactions with C—H bonds.

Apparently the reactivity of C—H bonds towards both d^8 species is determined mainly by steric factors and the strength of the M—C bonds formed, while the C—H bond energy is not of crucial importance. Perhaps the steric factors are less important for coordinatively unsaturated Ir(I) complexes than for Pt(II) complexes. E.g., Cp'IrPMe$_3$ reacts with the aromatic C—H bond of *p*-xylene [50], whereas Pt(II) species are not reactive towards *p*-substituted benzenes.

The remarkable lack of sensitivity of Pt(II) complex activity to the electronic density on Pt is apparently also a feature of the reactions of Ir(I) complexes. The Cp'Ir(I)L complexes are reactive towards alkanes with both L = PMe$_3$ and L = CO, which have very different donor properties. Taking into account that the electrophilic Ir(I) cationic complexes discussed above are also active toward C—H in alkanes, the situation might be comparable with that for Pt(II) complexes, where LPtCl$_3^-$ and L$_3$PtCl$^+$ are of similar reactivity.

V.4. The Activation of Alkanes by Low-Valent Complexes of Non-Platinum Metals

V.4.1. ZIEGLER-NATTA SYSTEMS

In 1969 in the first paper [1] on the homogeneous activation of alkanes by transition metal complexes the authors reported on the H—D exchange of methane with D$_2$ in the presence of the cobalt complex (Ph$_3$P)$_3$CoH$_3$. However, this work did not undergo any further development.

The choice of a solvent seemed to be a difficult problem for these reactions, since complexes of low-valent metals should not be stable in aqueous and other protic media, and hydrocarbon and other solvents containing C—H bonds should completely inhibit the alkane reaction.

Therefore, the results obtained more recently by Grigorjan *et al.* [52] were somewhat unexpected. It was shown that tetradeuteromethane is capable of exchanging deuterium atoms with hydrogen atoms of methyl groups in Ziegler-Natta catalytic systems, such as $TiCl_4 + AlMe_2Cl$; $(\pi\text{-}C_5H_5)_2TiCl_2 + AlMe_2Cl$; $VCl_3 + AlMe_2Cl$. The reaction was carried out in heptane solution at temperatures of $20-50°C$ and methane pressures of $0.3-4$ atm. As a result of H–D exchange in the gas phase, the appearance of methane molecules with a smaller number of deuterium atoms than at the start is observed, in particular, CH_2D_2 and CHD_3. The appearance of deuterium-substituted molecules, such as CH_2D_2, CHD_3, CDH_3, is also observed in the methane isolated by hydrolysis of the reaction products. The amount of CH_2D_2, CHD_3 and CH_4 in the gaseous reaction products increases with time.

The strong sensitivity of the results to traces of impurities has been reported. Pyridine, for instance, suppresses the exchange completely. The authors suggest a mechanism including a reversible stage of methyl group disproportionation in the coordination sphere of the transition metal atom in the complex, for example,

$$\begin{array}{c} \diagdown \quad\quad CH_3 \\ \diagup Ti \diagdown \\ \diagup \quad\quad CH_3 \end{array} \quad \rightleftharpoons \quad Ti{=}CH_2 + CH_4$$

In the presence of CD_4 such a process should result in the formation of methanes with different deuterium content. Although the reaction is carried out in a heptane solution, evidently the solvent does not prevent the reaction.

V.4.2. ADDITION OF METHANE TO UNSATURATED MOLECULES

In a later work [53] the authors discovered that *bis*-cyclopentadienylvanadium(II) catalyzes the H–D exchange of methane with molecular deuterium in benzene solution at $70°C$ which leads to deuterosubstituted methanes.

The yield of deuteromethanes reaches $2-6\%$ within 24 hours (a CH_4 content of 35% in initial mixture, pressure 250 Torr), the ratio of CH_3D, CH_2D_2, CHD_3 and CD_4 being $1:0.2:0.3:0.12$, respectively. In this case the multiple exchange in benzene molecules to give C_6D_6 and C_6HD_5 was found to proceed simultaneously with the H–D exchange between methane and deuterium. It was also found that $V(\pi\text{-}C_5H_5)_2$ catalyzes the H–D exchange of tetradeuteroethylene with methane, benzene molecules also participating (Table V.11).

The exchange is clearly multiple in nature with a large concentration of C_2H_4 and C_6D_6 molecules in the products and a significant yield of polydeuteromethanes (CD_3H). The mechanism proposed by the authors involves

TABLE V.11

Products of deuterium–hydrogen exchange in the system $Cp_2V + C_2D_4 + CH_4 + C_6H_6$
($70°C, P = 40$ atm., 100 h, $[Cp_2V] = 3.5 \times 10^{-2}$ M)

C_2D_4/CH_4	Products in moles per mole of Cp_2V								
	Ethylene				Methane			Benzene	
	C_2HD_3	$C_2H_2D_2$	C_2H_3D	C_2H_4	CHD_3	CH_2D_2	CH_3D	C_6D_6	C_6D_5H
0.88			0.17	33.46	0.49	2.43		30.16	4.06
0.34	0.13	0.84		10.68	0.64		1.6	6.34	1.46

a reversible H atom transfer from methane to olefin in the coordination sphere of vanadium:

This mechanism is possibly similar to that which may be responsible for the H atom transfer from cycloalkanes to 3,3-dimethylbutene molecules in cationic complexes of Ir (Section V.3). In effect, a further transfer of a hydrogen atom from a $-CH_3$ group could lead to the formation of ethane and a $>M=CH_2$ complex.

A recombination of alkyl groups in the vanadium coordination sphere (i.e. reductive elimination) would be another possible reaction path to yield propane, a product of methane addition to ethylene. Such a reaction has been recently discovered [54]. The formation of propane from methane and ethylene was shown to occur in the presence of a catalytic system, $Ti(OC_4H_9)_4 - Al(C_2H_5)_3$, at $20°C$ in benzene solution. The yield of propane amounts to 15% calculated on the amount of ethylene in the solution. If CD_4 is taken instead of CH_4, then the isotope-substituted molecule, $C_3H_4D_4$, is found to be the product of ethylene methanation.

The propane yield reaches, at present, only 20–25%, depending upon the titanium complex used. Nevertheless, the discovery of this new reaction is a significant step forward in the study of new homogeneous catalytic reactions of alkanes.

Addition of methane to acetylene in the presence of the homogeneous system $Fe(acac)_3$ and $AlEt_3$ leads to the formation of some propylene (up to *ca.* 3% per $Fe(acac)_3$) [54]. Propylene-d_4 is detected when CD_4 is taken instead of CH_4.

The reactions mentioned are carried out in a hydrocarbon solvent, such as heptane or benzene, which is present in a very great excess over the reacting methane. Methane is usually the least reactive of alkanes, which themselves are less reactive than benzene, so it might be expected that the products of the reaction with methane should constitute only a small fraction of the total reaction products. In effect, methane is evidently responsible for a large portion of the products (e.g., in H–D exchange with D_2, $C_2 D_4$, etc.) showing that it is very reactive as compared with other C—H-containing molecules present.

Does this mean that the small size of atoms of the first transition metal series leads to very serious steric hindrance, to which methane should be least sensitive, or that in these systems, which have strongly pronounced donor properties, the reactivity of the alkane series has a maximum value for methane? Hopefully, further studies will clarify these questions.

V.4.3. DEHYDROGENATION OF ALKANES IN THE PRESENCE OF RHENIUM HYDRIDE COMPLEX [55]

Dehydrogenation of cycloalkanes in the coordination sphere of metal complexes, found first for a cationic Ir(I) complex (Section V.3.2), was also demonstrated for the rhenium complexes $L_2 ReH_7$ (L = PPh_3 or $PEt_2 Ph$). These complexes were shown to dehydrogenate cyclopentane to form $L_2 (\eta^5 -C_5 H_5)ReH_2$ in the presence of 3,3-dimethyl-1-butene. The reaction in cyclopentane solution proceeds at 80° according to the scheme

Later the authors [55] found that linear alkanes could be also dehydrogenated in this system to give diene complexes, e.g.,

The diene in this complex can be converted further to monoolefin in the presence of trimethyl phosphite. E.g., n-pentane can be selectively converted in 1-pentene according to the scheme

Ar = p-MeC$_6$H$_4$ and Ph.

The authors [55] suggest that an alkane molecule reacts with an intermediate $L_2 ReH_3$, the interaction proceeding initially as an oxidative addition

$$L_2 ReH_3 + RH \rightleftharpoons L_2 \overset{\overset{\textstyle R}{\textstyle |}}{ReH_4}$$

This recent work demonstrated once more the potential possibilities for functionalization of alkanes in the presence of homogeneous coordination catalysts.

V.5. A Comparison of Homogeneous and Heterogeneous Activation of Hydrocarbons

Transition metal complexes, in their reactions with alkanes, show some similarity in certain aspects with all the other species active towards alkanes, such as free metal and non-metal atoms, radicals, ions, as well as carbenes, strong electrophiles or superacids. But probably the most important for chemistry is their analogy to heterogeneous catalysts reacting with alkanes on a surface.

At present similarity between mechanisms of homogeneous and heterogeneous catalysis is generally recognized. In both the cases the interaction occurs in the metal coordination sphere of the catalyst. The equilibrium concentration of the active centers in solutions may be considerably below that of the stable complexes present. Similarly, the surface concentration of active centers may be lower than would correspond to the surface area, since not all the surface atoms are active.

A complete comparison of homogeneous and heterogeneous activation of alkanes is still difficult to make, since the number of examples of homogeneous activation of alkanes, though rapidly increasing, is still far fewer than those of heterogeneous catalytic systems. This is mainly due to an objective factor; the reactive chemical inertness of alkanes. In the case of heterogeneous systems containing, for example, metallic surfaces, this inertness is easy to overcome by carrying out the reaction at high temperatures (up to several hundred degrees). This is usually difficult to achieve for the complexes in solution, which are usually unstable at high temperatures. Therefore, only a few cases, where comparison seems to be plausible, will be considered. As we have seen, platinum(II) complexes in aqueous solution appeared to be more active towards the H–D exchange of hydrocarbons than the analogous complexes of other platinum metals. Metallic platinum usually is also a more active catalyst of H–D exchange (with D_2 and D_2O) than the other platinum metals. As in the case of Pt(II) complexes, it is probably due to the comparatively strong Pt–C bond which is the driving force of the reaction. In the case of metallic platinum it is confirmed by the facts that the Pt surface which is in direct contact with organic substances is covered with a carbonacious film.

The steric hindrances in reactions in solution, which are due mainly to the shielding effect of ligands at the central metal ion, are expected to be less significant in heterogeneous activation. In effect, in heterogeneous catalysis, the selectivity is generally determined by the relative C—H bond strength in alkanes and the rate increases in branched alkanes in the order $1° < 2° < 3°$. For example, in case of metallic nickel, the selectivity corresponds to a ratio $1° : 2° : 3° = 1.30 : 90$. For such metals as Pd, Rh, Ir and Pt, the difference between the activities of primary and secondary C—H bonds is small. However, in all the known cases, a tertiary C—H bond is exchanged more rapidly than a secondary and a primary bond, in sharp contrast to homogeneous exchange at platinum complexes, where the H atom at a tertiary carbon atom is practically not exchanged at all. For example, in the case of multiple exchange at a Rh surface, isobutane in the initial stages forms considerable quantities of d_{10} hydrocarbon, whereas isobutane-d_{10} is practically absent in the multiple exchange of iso-C_4H_{10} in solutions of Pt(II) complexes. Garnett [56] has made a special comparison of the H—D exchange of aromatic hydrocarbons with D_2O in homogeneous solutions of Pt(II) complexes and on the surface of platinum black formed from PtO_2 in reactions with borohydride. A general similarity between homogeneous and heterogeneous exchange is observed. For example, in both cases, naphthalene and anthracene initially exchange only β-hydrogen atoms in these molecules, which indicates the formation of β-aryl platinum derivatives both in solution and at the surface.

For substituted benzenes, in both cases only m- and p-hydrogen atoms with respect to the substituent in the benzene ring are exchanged. Marked distinctions were observed only for o-terphenyl [57].

In homogeneous exchange with the participation of $PtCl_4^{2-}$, the usual inhibition of the H—D exchange was observed in the o-position to the substituent, particularly in the case of X atoms of the central ring. In heterogeneous exchange, an H atom in the X position is readily exchanged with D_2O, demonstrating that steric hindrance is indeed less significant in this particular case.

Multiple exchange is widespread both in homogeneous and heterogeneous catalyses of isotope exchange. However, the mechanisms of homogeneous and heterogeneous multiple exchange have some differences related to the nature of the active centers. As we have seen, a homogeneous exchange of alkanes involves an intermediate formation of carbene complexes as a general case. The probability of formation of the carbene is very close for methane and ethane: $k_{11}/k_{-1} = 4.9$ for CH_4 and 4.8 for C_2H_6 (Scheme V.7, Section V.2.4). In heterogeneous exchange, the mechanism with the intermediate formation of 1,1-bonded complexes is found rarely: only in the cases of such molecules as methane, neopentane, etc. For ethane, 1,1-bonded complexes are observed for neither of the systems studied. That is why a stepwise multiple exchange in ethane molecules, which was observed in the exchange of C_2H_6 in homogeneous aqueous solutions in the presence of Pt(II) complexes, was not found for various cases of heterogeneous catalytic exchange. 1,2-Bonded complexes in heterogeneous catalysis may be thought to form with greater ease due to the possibility of the participation of other platinum atoms located adjacent to the active center.

$$
\begin{array}{ccc}
\mathrm{CH_2\!-\!CH_3} & & \mathrm{CH_2\!-\!CH_2} \\
| \qquad\quad & & | \qquad\quad | \\
\sim\!\!\mathrm{Pt} \sim \mathrm{Pt} & \longrightarrow & \sim \mathrm{Pt} \sim \mathrm{Pt\!-\!H}
\end{array}
$$

It is possible that the formation of carbene complexes of Pt(II) in aqueous solution is connected with the Pt(II) polar effect on the adjacent C—H bond which promotes the H^+ elimination.

Nevertheless, the differences with respect to 1,1- and 1,2-bonding of the intermediate species are quantitative. The heterogeneous catalytic exchange of methane and neopentane, which is multiple in nature, apparently includes the formation of metal carbene complexes. It is also noteworthy that, along with 1,1- and 1,2-bonded intermediate species, as observed for homogeneous exchange, in the exchange of hydrocarbon with D_2 at a metal surface, 1,3-bonded intermediate species are also formed, which are more stable for platinum than for other metals.

Hence, we may conclude that, together with extensive analogies in the homogeneous and heterogeneous activation of alkanes, there are also some important differences. A more detailed comparison of these two types of reactions of alkanes, particularly for different metals, will require a larger quantity of data on the homogeneous activation of hydrocarbons.

References

1. N. F. Goldshleger, M. B. Tyabin, A. E. Shilov, and A. A. Shteinman: *Zh. Fiz. Khim.*, **43**, 2174 (1969).
2. a. A. E. Shilov and A. A. Shteinman: *Coord. Chem. Rev.*, **24**, 97 (1977);
   b. D. E. Webster: *Adv. in Organomet. Chem.*, (Eds. F. G. A. Stone and R. West), V. 15, p. 147, Academic Press, New York (1977).
3. J. L. Garnett and R. J. Hodges: *J. Amer. Chem. Soc.*, **89**, 4546 (1967).
4. R. J. Hodges and J. L. Garnett: *J. Phys. Chem.*, **72**, 1673 (1968).
5. A. C. Skapski and F. A. Stephens: *J. Chem. Soc., Chem. Commun.*, 1008 (1969).
6. M. B. Tyabin, A. E. Shilov, and A. A. Shteinman: *Dokl. Akad. Nauk SSSR*, **198**, 380 (1971).
7. A. A. Shteinman: *Thesis*, Chernogolovka (1978).
8. N. F. Goldshleger and A. A. Shteinman: *React. Kinet. Catal. Lett.*, **6**, 43 (1977).
9. C. H. Langford and H. B. Gray: *Ligand Substitution Processes*, W. A. Benjamin, Inc., New York, Amsterdam (1965).
10. Ya. D. Fridman: *Koord. Khim.*, **1**, 1155 (1975).
11. D. M. Adams, J. Chatt, and B. L. Shaw: *J. Chem. Soc.*, 2047 (1960).
12. J. Chatt and B. L. Shaw: *J. Chem. Soc.*, 5075 (1962).
13. R. J. Hodges, D. E. Webster, and P. B. Wells: *J. Chem. Soc. A*, 3230 (1971).
14. J. L. Garnett and J. C. West: *Aust. J. Chem.*, **27**, 129 (1974).
15. R. J. Hodges, D. E. Webster, and P. B. Wells: *J. Chem. Soc., Dalton*, 2577 (1972).
16. L. F. Repka and A. A. Shteinman: *Kinet. Katal.*, **15**, 805 (1974).
17. R. R. Shrock: *J. Amer. Chem. Soc.*, **97**, 6578 (1975).
18. E. S. Rudakov and A. A. Shteinman: *Kinet. Katal.*, **14**, 1346 (1973).
19. N. F. Goldshleger, I. I. Moiseev, M. L. Khidekel, and A. A. Shteinman: *Dokl. Akad. Nauk SSSR*, **206**, 106 (1972).
20. E. S. Rudakov: *Dokl. Akad. Nauk SSSR:* **229**, 149 (1976).
21. K. G. Powell and F. J. McQuillin: *Tetrahedron Lett.*, 2213 (1971).
22. E. S. Rudakov, V. P. Tretyakov, S. A. Mitchenko, and A. V. Bogdanov: *Dokl. Akad. Nauk SSSR*, **259**, 899 (1981).
23. N. F. Goldshleger, V. V. Eskova, A. E. Shilov, and A. A. Shteinman: *Zh. Fiz. Khim.*, **46**, 1353 (1972).
24. V. V. Eskova, A. E. Shilov, and A. A. Shteinman: *Kinet. Katal.*, **13**, 534 (1972).
25. J. L. Garnett and J. C. West: *Syn. Inorg. Metal-Org. Chem.*, **2**, 25 (1972).
26. J. R. Sanders, D. E. Webster, and P. B. Wells: *J. Chem. Soc., Dalton*, 1191 (1975).
27. N. F. Goldshleger, V. V. Lavrushko, A. P. Khrushch, and A. A. Shteinman: *Izv. Akad. Nauk SSSR, ser. khim.*, 2174 (1976).
28. Yu. V. Geletii and A. E. Shilov: *Kinet. Katal.*, **24**, p. 486 (1983).
29. V. P. Tretyakov, G. P. Zimtseva, E. S. Rudakov, and A. N. Osetskii: *React. Kinet. Catal. Lett.*, **12**, 543 (1979).
30. V. P. Tretyakov, E. S. Rudakov, A. A. Galenin, and R. I. Rudakova: *Dokl. Akad. Nauk SSSR*, **225**, 583 (1975).
31. E. S. Rudakov, V. P. Tretyakov, A. A. Galenin, and G. P. Zimtseva: *Dopov. Akad. Nauk Ukr. RSR, ser. khim. B*, No. 2, 148 (1977).
32. V. P. Tretyakov, E. S. Rudakov, A. V. Bogdanov, G. P. Zimtseva, L. I. Kozhevina: *Dokl. Akad. Nauk SSSR*, **249**, 878 (1979).
33. V. V. Lavrushko, A. E. Shilov, and A. A. Shteinman: *Kinet. Katal.*, **16**, 1479 (1975).

34. V. P. Tretyakov, E. S. Rudakov, R. I. Rudakova, A. A. Galenin, and G. P. Zimtseva: *Zh. Fiz. Khim.*, **51**, 1020 (1977).

35. N. F. Goldshleger, V. M. Nekipelov, A. T. Nikitaev, K. I. Zamaraev, A. E. Shilov, and A. A. Shteinman: *Kinet. Katal.*, **20**, 538 (1979).

36. G. B. Shul'pin, L. P. Rosenberg, R. P. Shibaeva, and A. E. Shilov: *Kinet. Katal.*, **20**, 1570 (1979);
    G. B. Shul'pin: *ibid*, **22**, 520 (1981);
    G. B. Shul'pin and G. V. Nisova: *ibid*, **22**, 1061 (1981).

37. G. B. Shul'pin, A. E. Shilov, A. N. Kitaigorodskii, and J. V. Zeile-Krevor: *J. Organomet. Chem.*, **201**, 319 (1980).

38. R. P. Shibaeva, L. P. Rozenberg, R. M. Lobkovskaya, A. E. Shilov, and G. B. Shul'pin: *J. Organomet. Chem.*, **220**, 271 (1981).

39. V. V. Lavrushko, S. A. Lermontov, and A. E. Shilov: *React. Kinet. Catal. Lett.*, **15**, 269 (1980).

40. L. A. Kushch, V. V. Lavrushko, Yu. S. Misharin, A. P. Moravskii, and A. E. Shilov: *Nouv. J. Chim.*, 7, No. 12 (1983).

41. V. V. Zamashchikov, E. S. Rudakov, S. A. Mitchenko, S. L. Litvinenko: *Teoret. Exp. Khim.*, No. 4, 510 (1982).

42. I. I. Moiseev: *Zh. Vses. Khim. Ob-va im. Mendeleeva*, **22**, 30 (1977).

43. A. F. Shestakov: *Koord. Khim.*, 6, 189 (1980).

44. A. F. Shestakov and S. M. Vinogradova: *Koord. Khim.*, 9, 248 (1983).

45. J. L. Garnett, M. A. Long, A. B. McLaren, and K. B. Peterson: *J. Chem. Soc., Chem. Commun.*, 749 (1973).

46. J. L. Garnett, M. A. Long, and K. B. Peterson: *Aust. J. Chem.*, **27**, 1823 (1974).

47. M. R. Blake, J. L. Garnett, I. K. Gregor, W. Hannan, K. Hoa, and M. A. Long: *J. Chem. Soc., Chem. Commun.*, 930 (1975).

48. R. H. Crabtree, J. M. Mihelcic, and J. M. Quirre: *J. Amer. Chem. Soc.*, **101**, 7738 (1979).

49. R. H. Crabtree: *Acc. Chem. Res.*, **12**, 331 (1979).

50. A. H. Janowicz and R. G. Bergman: *J. Amer. Chem. Soc.*, **104**, 352 (1982);
    idem: ibid, **105**, 2929 (1983).

51. J. K. Hoyano and W. A. G. Graham: *J. Amer. Chem. Soc.*, **104**, 3723 (1982).

52. E. A. Grigorjan, F. S. Dyachkovskii, and I. R. Mullagaliev: *Dokl. Akad. Nauk SSSR*, **224**, 859 (1975).

53. E. A. Grigorjan, F. S. Dyachkovskii, S. Ya. Zhuk, and L. I. Vyshinskaja: *Kinet. Katal.*, **19**, 1063 (1978).

54. a. N. S. Enikolopjan, Kh. R. Gyulumjan, and E. A. Grigorjan: *Dokl. Akad. Nauk SSSR*, **249**, 1380 (1979);
    b. E. A. Grigorjan, Kh. R. Gyulumjan, E. I. Gurtovaya, N. S. Enikolopjan, and M. A. Ter-Kazarova: *Dokl. Akad. Nauk SSSR*, **257**, 364 (1981).

55. D. Baudry, M. Ephritikhine, and H. Felkin: *J. Chem. Soc., Chem. Commun.*, 1243 (1980);
    idem: ibid, 606 (1982);
    idem: ibid, 1235 (1982).

56. K. P. Davis and J. L. Garnett: *Aust. J. Chem.*, **28**, 1699 (1975).

57. K. P. Davis and J. L. Garnett: *J. Chem. Soc., Chem. Commun.*, 79 (1975);
    K. P. Davis and J. L. Garnett: *Aust. J. Chem.*, **28**, 1713 (1975).

CONCLUSION

The experimental data given in this book show that metal complexes in high, medium, and low oxidation states exhibit an appreciable reactivity towards covalent C—H bonds, among them C—H bonds in alkanes. The most significant results, particularly for the reactions of the complexes in medium and low oxidation states, have been obtained during the last decade. There is no doubt that this field will continue to be extensively developed in the coming years.

The reactions of alkanes in the metal coordination sphere justified the hopes set on them, demonstrating, for example, an unusual selectivity and high reactivity for hydrocarbons (such as methane and ethane) which were traditionally thought to be inert.

The reactions known are as yet interesting only from the point of view of basic knowledge (with the exception of earlier discovered reactions of more traditional oxidation, in which the metal complexes usually participate in the radical-chain process). However, it is expected that in the near future the reactions of activation of C—H bonds in alkanes by metal complexes will be utilized in industrially important processes. This achievement, in its turn, will attract the attention of investigators studying the coordination catalysis to the field considered and will contribute to its development.

(a. Shilov)

INDEX

acetic acid
  oxidation in, 14, 23–4, 87, 164, 169, 171
  as solvent, 81, 145, 158, 163, 182
acetic anhydride, 124
acetonitrile, 25, 58, 67, 116, 121
  oxidation in, 108–9
acetoxylation, 14, 23
acidic interaction, 5
acrylonitrile, 58
'activated' aliphatic C H bonds, 20–32
activation energy, 170, 180
activation of C-H bonds
  by polar substituents, 24–6
activation energy, 142, 170, 180
adamantane, 36, 38, 69
aliphatic hydrocarbons, 9, 17
  hydroxylation, 102, 123
alkanes
  acid-base interaction, 5
  acidities, 1, 2
  activation by low-valent complexes of non-platinum metals, 185–9
  activation by metal complexes, 141–91
  activation on surface of metal and metal oxides, 49–50
  characteristics, 1, 2
  C-H bond energies, 1, 2, 9, 20
  chemical inertness, 1
  chlorination, 38
  dehydrogenation by Ir(I) complexes, 182–4
  dehydrogenation in presence of rhenium hybride complex, 188–9
  deuteration, 151, 159
  electron affinities, 1, 2
  electrophile addition, 5
  hydroxylation, 45–7, 108, 123, 125
  ionization potentials, 1–3
  isomerization, 39
  H-D exchange, 146–50, 181–5
    nitration, 38
    nitrolysis, 38
    oxidation by catalyst, 3
    oxidation by compounds of cobalt(III), 67–9
    oxidation by metal compounds, 62–73
    oxidation in presence of Pt(II) and Pt(IV) complexes, 162–4
    oxidation in sulfuric acid, 70–3
    proton affinities, 1, 2
    protonation, 5, 35
    reactions, comparative, 10
    reactions with atoms, 40–3
    reactions with carbenes, 43
    reactions with electrophiles, 35–9
    reactions with electrophilic oxidants, 125–36
    reactions with free radicals, 40–3
    reactions with metal atoms and ions, 58–60
    reactions with Pt(II) complexes, 9, 144–80
    reactions with transition metal complexes, 6
alkenes, 101
alkylaromatic compounds, 20, 21
  catalytic oxidation, 23, 86
  reactions in presence of platinum II complexes, 23
alkylaromatic hydrocarbons, 51, 67
  oxidation, 82
alkylation, 37, 39
alkylbenzenes, 15, 21, 143
  deuteration, 151
alkyl-carbenium, 39
alkyl hydride, 6, 9, 18, 142
  complexes, 143, 144, 179
alkyl metal derivatives, 155, 163
alkyl methacrylate, 19
alkyloysis, 37
alkylpalladium derivatives, 16

197

alkyl platinum derivatives, 156, 157, 167, 169
alkyl platinum complexes, 161, 162, 171–7, 178
aluminium isopropoxide, 51
aluminium oxides, 50
aminium radicals, 41–3
amino-oxides, 42
ammonia, 58, 172
ammonium salts, 172
ammoxidation, 58
anisole, 173
anthracene, 190
aquo-complexes, 149
arenes
    C-H bond cleavage, 11, 141
    ionization potential, 3
    isotope exchange of, with deterium, 19–20
    oxidation, 119
    oxidative coupling, 5, 14
    reactions in platinum salt solutions, 17
    reactions in presence of platinum (II) complexes, 144–80
    reactions with electrophilic oxidants, 11–12
    reactions with low oxidation state metal complexes, 17–19
π-arene complex, 19
arene hydrocarbons, 5
aromatic amines, 12
aromatic compounds, 167
    hydroxylation, 106
    ionization potential, 20
    oxidation, 17
    'platination', 177
aromatic hydrocarbons, 5, 9, 10, 11–20, 51, 141, 144
    H-D exchange, 144, 146–50, 190
    hydroxylation, 102, 123, 125
    oxidation, 57
    reaction with cobalt(III) ions, 67
    redox potentials, 128
aromatic molecules
    oxidation by bivalent palladium, 14
    oxidative addition, 18
    oxidative coupling, 14
aryl derivatives, 17, 143
σ-aryl metal derivatives, 15

arylmetal hydrides, 17
arylpalladium derivatives, 15
arylphosphite complexes, 28
aryl platinum complexes, 171–7
β-aryl platinum derivatives, 190
ascorbic acid, 106-8
auration, 14
azobenzene, 27

benzene
    activation energy, 51
    C-H bond energy, 1
    H-D exchange, 17, 50, 150, 162–4
    hydroxylation, 101
    multiple exchange, 155, 182
    oxidation, 12, 15, 67, 69, 87, 167, 170–3
    reactivity, 185, 188
benzene acetate, 23
benzene ring, 13, 27, 51, 65, 153, 182, 190
    chlorination, 22
benzyl acetate, 23
benzyl radicals, 21
biaryls, 17
bicyclooctene, 183
biological oxidation, 160
    of alkanes, 43, 88–103, 135, 136, 160
    of hydrocarbons, 103–4, 119–20, 123
    mechanism of, 101–3
    steric hindrance, 8
biological hydroxylation, 123
borohydride, 190
branching chain mechanism, 114–15, 116
bronbobenzene, 151
bromide, 82, 83, 86
bromine, 82, 83, 86
butane, 44, 87
    isomerization, 39
    oxidation, 87
t-butyl, 12, 42
n- butylphosphines, 30

carbanions, 50
carbenes, 43–5, 54, 97, 158, 161, 162
carbene complexes, 156, 191
carbene-platinum complex, 162
carbon, 165–7
carbonium, 35, 37–9
catalyst regeneration, 3

chain initiation, 75, 77–8
chemiluminescence, 81, 83, 85
chemisorption, 49, 53
chloride, 164, 170
chloride complexes, 169
chlorination, 42, 43, 164, 169
chloroacetic acid, 163, 164
chloroalkanes, 163, 164
chloroanisole, 173
chlorobenzene, 151, 161, 162, 174
chloropentanes, 42
chlorosubstituted arenes, 17
chromic acid, 63, 64, 67
chromium oxide, 50
chromyl chloride, 63
cobalt(III), 21, 67–9, 78, 82, 86, 87
cobalt-bromide catalysis, 82
cobalt stearate, 77, 79
π-complexes, 155, 158
co-condensation, 59
coordination catalysis, 6
copper, 100
copper chloride, 164
copper dichloride, 112
coulomb interaction, 4
cracking, 40, 49, 51, 55
cycloalkanes, 50, 51, 53, 65, 183
cyclodecane, 185
cyclododecanone, 43
cycloheptane, 153
cyclohexane
    cooxidation, 121
    dehydrogenation, 164
    N-D exchange, 146, 148–51, 161
    hydroxylation, 93, 101, 106, 109, 111
    isotope exchange, 50
    oxidation, 58, 65, 71, 87, 116–19, 122, 124, 167
cyclohexanol
    chlorination, 43
    hydroxylation, 101, 121
    oxidation, 69, 111, 118, 122, 124
cyclohexanone, 111, 184
cyclohexene, 50, 67
cyclometallation, 26–7, 28, 31
cyclooctadiene, 182
cyclooctane, 153, 182, 185
cyclooctene, 182, 183
cyclopentadecanone, 43

cyclopentadienyl, 19, 183
cyclopentane, 53, 71, 183, 188
cyclopropane, 50, 160, 185
cytochrome P-450, 92–9, 101–3, 119, 123–5, 166

decane, 164
'degenerate' chain branching, 74, 78, 127
dehydrochlorination, 164
dehydrocyclization, 49, 55–7
dehydrogenases, 89
dehydrogenation, 55–7
    of alkanes by Ir(I) complexes, 182–4
    of alkanes in presence of Pt(II) and Pt(IV) complexes, 162–4
    of alkanes in presence of rhenium hydride complex, 188
    catalytic, 49, 55
    oxidative, 93
deuterated benzene, 20
deuterated olefins, 16
deuterium
    isotope exchange, 19–20, 51, 142
    molecular, 54, 143, 186
    multiple exchange, 155, 156
deuteroalkanes, 157
deuterocarbon, 156
deuteroethanes, 157, 158
deuterohydrocarbons, 156, 157
deuteromethanes, 101, 156, 157, 186
deuterophenyl deuteride, 184
deuteropropanes, 159, 160
dialkyl complexes, 7
dianisyl, 173
diaryls, 14–16
diarylmercury compounds, 15
diarylmethanes, 15
dichlorocarbene, 45
dichloroethane, 183, 184
dicyclopentadienyltungsten, 18
dideuterium, 36, 51
diene complexes, 188
diene molecules, 183
dihydrogen, 1, 2, 9, 36, 39, 142
dihydroiridium complexes, 26
3,3–dimethylbutene, 183, 188
3,3–dimethylpentane, 53
2,2–dimethylpropane, 166
dimethyl sulfoxide, 149

dioxygen
  cooxidation by, 121
  oxidation by, 75, 77–8, 85, 87–9, 91,
    113, 116, 118, 120
  in oxidation of Cytochrome-450, 95, 97
  redox potential, 13
dioxygenases, 89–91, 102
diphenyl, 15, 162, 171
1,1–diphenylethylene, 17
diphenyl-n-propylphosphine, 30
durene, 15

electron acceptors, 21, 127, 154
electronegative substituents, 25, 41
electron transfer, 3, 21, 82, 86, 89, 126–
    30
electrophiles, 147
electrophilic chlorination, 38
electrophilic mercuration, 177
electrophilic oxidants, 125–36
electrophilic oxygenation, 38
electrophilic proton replacement, 17
electrophilic reagents, 36
electrophilic substitution, 12, 13, 16, 23,
    27, 149, 177
  in alkenes, 35, 36, 127, 135, 136
electrostatic interaction, 4
α-elimination, 29
β-elimination, 29, 52
endothermic process, 129, 133, 136
enzymes, 88–92, 96–101, 104, 106
etard reaction, 63
ethane, 44, 143, 153, 155, 157, 166
  multiple exchange, 156, 158
ethyl acetate, 25
ethylene, 1, 20, 187
ethylidene complex, 158
Evance-Polanyi rule, 40
exothermic process, 104, 130, 136

fluorine, 131
m-fluoroazobenzene, 28
free-radicals, 4, 21, 41, 58, 82, 104
  scavengers of, 45

gold, 55, 57

halogenation, 40
Hammett equation, 12, 72

'hard hydrogen bond', 32
'heat stable factor', 94–5
heptane, 68, 87, 188
heterogeneous activation, 189–90
heterogeneous catalysis, 49, 57, 191
heterogeneous oxidation, 57–8
heteropolyacid, 164
hexane, 164
histidine, 96
homogeneous activation, 49, 189–91
homogeneous catalysis, 191
Hückel method, 180
hybridized carbon atom, 10
hydrazine hydrate, 171, 174
hydride-ion abstraction, 134–5
hydroalkylmetal complexes, 184
hydrocarbons
  branched, 40
  homogeneous and heterogeneous activa-
    tion, 189–91
  hydroxylation, 106–13
  ionization potential, 152, 153
  oxidation, 62, 74–8, 103–6
  radicals, 117
  reactions with Pt(II) complexes, 177
  reactivity, 164–70
  stoichiometric oxidation, 62
hydrocarbon radicals, 4
hydrocracking, 52
hydrogenation catalyst, 50
hydrogen halides, 183
hydrogenolysis, 55, 57
hydrogen peroxide, 115, 116, 118
hydrolysis, 186
hydroperoxides
  olefin epoxidation by, 122
  in oxidation, 74–85, 97, 117–118
  radicals, 117
hydroxilation, 42, 46–7, 88, 90–5, 98
  enzymatic, 97, 103
  of hydrocarbons, 102, 106–13
  photochemical, 108
hydroxyl radicals, 117

indigo, 105
iodosobenzene, 121–5
iridium complexes, 19, 25, 32
  in oxidation, 65–7, 144, 183–5
iron complexes, 18, 32, 91, 119, 123

iron porphyrin complexes, 124
irradiation, 59
isobutane, 44, 87, 160, 167, 190
isomers, 13
isomerization, 49, 57, 158
isooctane, 160
isotope exchange, 50-4, 55, 142, 145, 150

ketogluterate, 102
ketones, 123
kinetic isotope effect, 58

Lewis acid, 183

manganese, 32
mercuration, 5, 12-14
mercuric acetate, 12, 13
mercuric perchlorate, 12, 13
mesitylene, 15, 24
metal complexes
    activation of S-H bonds in ligands,
        26-32
    coupled oxidation, 103
    electrophilic, 5
    high valency, 5, 8
    low valency, 9, 10
    oxidation, 3, 5, 7, 9, 106-13, 121,
        144
    reactions with alkanes, 3, 4, 49-60
    reaction with alkylaromatic compounds,
        20
    reaction with benzene, 11
    reactions with compounds containing
        'activated' C-H bonds, 11-32
metal coordination catalysis, 4
metallocycles, 26-7, 29-30, 160, 161
metallation, 173
o-metallation, 27, 28
metallic platinum, 51, 145, 189-90
metallic sodium, 18
metal oxides, 50
methane, 131, 143, 153, 157
    activation energy, 51
    addition to unsaturated molecules, 186-
        8
    C-H bond, 2, 9, 20, 166
    H-D exchange, 161
    heterogeneous catalytic exchange, 191
    multiple exchange, 54, 158

oxidation, 88, 99, 101, 164, 174, 175
protonation, 36
reaction enthalpy, 9
reaction with methylene, 44, 45
reaction products, 188
reaction rate, 58, 171
methanemonooxygenase, 99-101, 102,
    103, 136, 166
methanol, 25, 109, 164, 174
methine complex, 161
methonium, 35
methyl chloride, 38, 174
methyl cyanoacetate, 25
methylene, 32, 43, 44, 45, 150, 165
methylene chloride, 184
methylene-platinum complex, 156, 174
methyl hydrogen, 31
methylplatinum, 177
methyl-tantalum complex, 156
microscopic reversibility, principle of, 7
molybdenum peroxides, 122, 123
monoalkyl-substituted benzene, 15
monodeuteromethane, 143
monoxygenases
    in biological oxidation, 89-95, 98-101
    chemical models, 103-6
    in oxidation, 107, 118, 125, 130, 136,
        166
monosubstituted arenes, 15
monosubstituted benzenes, 13
Mössbauer spectroscopy, 59
multiple exchange, 52, 54, 155-62, 191

naphtalene, 18, 31, 172, 190
    H-D exchange, 151, 160
    hydroxylation, 123
    multiple exchange, 155
naphthyl, 172
naphthyl hydride, 24
β-naphthyl hydride, 31
β-naphthylruthenium hydride, 18
neopentane, 37, 51, 53, 103, 184, 185, 191
niobium, 19
nitrobenzene, 12
o-nitrotoluene, 172, 173
nucleophiles, 147
nucleophilic substitution, 36

octane, 164, 166

olefins, 5, 29, 30, 53, 144, 164, 182
  arylation, 16
  π-bound, 54
  catalytic hydrogenation, 7, 55, 183
  catalytic oxidation, 16
  C-H bond cleavage, 19
  epoxidation, 102, 107, 119, 122–5
  hydrogenation, 50
  isonerization, 51
  oxidation, 45, 57, 119
  oxidative coupling, 14, 15
orbital interaction, 4, 5
osmium complex, 18
oxenoid mechanism, 107, 119, 135
oxidases, 89
oxidation
  of alkanes by metal compounds, 62–73
  of alkanes in presence of Pt(II) and Pt(IV) complexes, 162–4
  of aromatic compounds, 17
  chain mechanism, 74–6
  chain reactions, 77–8
  deep, 57
  enzymatic, 88–90
  heterogeneous, 57–8
  of hydrocarbonds, 74–8, 77–88
  mechanism of, 164–70
  of methane, 99–101
  of olefins, 45, 57, 119
  two-electron, 133–6
oxidative coupling, 14, 15, 23
oxidative dimerization, 17
'oxidative homolysis', 126, 131
oxygen, molecular, 119–20, 127, 130, 145
oxygenases, 89, 121–5

palladium, 14, 23, 27
  complexes, 15, 16, 32, 73
  isotope exchange, 53
palladium black, 23–4
Pauling formula, 9
pentane, 42, 44, 160, 166, 185
n-pentane, 93
n-pentyl hybride, 184
peracids, 45
perchloric acid, 12, 13
periodic system, 7
permanganate, 24

peroxidases, 99
peroxides, 88, 121–5
peroxide theory, 74–6, 105
perylene, 128
phenanthrene, 153, 155
phenols, 11
phenyl, 12
phenylcarbene, 45
phenylhydrazine, 112
phosphatidyl choline, 95
phosphine, 28
phosphine methyl, 18
platinum, 55, 57, 148, 169, 177, 191
  alkylation, 159, 160
  alkyl derivatives, 143
  aryl derivatives, 143
  metallic, 144
  tetravalent, 179
platinum black, 190
platinum complexes, 9, 17, 19, 29
  in dehydrogenation of alkanes, 162–4
  in H-D exchange, 144–54
  mononuclear, 55
  in multiple exchange, 155–62
  in oxidation of alkanes, 162–4
  reactions with arenes and alkanes, 136, 143, 144–80, 182
platinum halides, 171–7
platinum salt solutions, 17
plumbation, 5, 14, 177
Polanyi–Semenov rule, 40
polar substituents, 24
polydeuteromethanes, 186
polyhydrides, 19
porphyrin, 96, 98, 119, 123
  'picket fence', 119
potassium permanganate, 64
Prilezhaev reaction, 46
propane, 164, 166, 185
  activation energy, 51, 57
  dehydrogenation, 55
propulene, 187
protic media, 142
proton elimination, 3, 5, 126, 128–30, 178
proton solvation, 3
proton substitution, 5, 16, 36
pyrene, 145, 153, 182
pyridine, 12, 186

redox potential, 97, 127–30, 133
   alkanes, 3, 62, 63
      bromine, 86
      iridium, 66
      iron, 103
      stannous chloride, 113
resonance integral, 4
rhenium, 144
rhenium hydride complex, 188–9
rhodium, 19, 32, 53, 54, 144
rhodium complexes, 182
rhodium trichloride, 182
'rolling over' process, 53
ruthenium complexes, 18, 65–7
ruthenium naphthylhydride complex, 30

scavengers, 45
1,2-shift, 1,3-shift, 1,4-shift, 160
silver nitrate, 15
singlet methylene, 43, 44, 179
singlet molecules, 43
sodium acetate, 15, 23, 82
sodium borohydride, 174
'soft hydrogen bond', 32
stannous chloride, 110, 113–16
stepwise distribution, 157, 158
stepwise exchange, 54
stereoselectivity, 121, 122, 124, 136
steric hindrance
   in activation of hydrocarbons, 190
   in biological oxidation, 8
   in dehydrogenation of alkanes, 183, 188
   in H-D exchange, 150, 154
   in isotope exchange, 51
   in mercuration, 12
   in multiple exchange, 160, 161
   in oxidation, 69, 72, 86, 101, 119
   in oxidative coupling, 15
   in 'oxidative homolysis', 132
   in reactions between alkanes and electrophiles, 36–7
   in reactions of metal complexes, 2
   in two-electron oxidation, 136
stoichiometry, 118, 119
styrene, 17, 101
substituted arenes, 15
sulfuric acid, 70–3, 131
superacids, 5, 35–7, 45
surface diagnostic techniques, 57

Taft equation, 72, 153
tantalum, 19, 29, 156
terephthalic acid, 23
tetradeuteroethylene, 186
tetradeuteromethane, 186
2,2,3,3-tetramethylbutane, 166
tetramethylsilane, 26
thallation, 14, 177
tin, 116, 117

titanium, 31, 32, 144, 187
toluene, 18, 20, 50, 57, 67, 173, 174
   conversion to benzaldehyde, 63
   hydroxylation, 112
   oxidation, 22, 82, 101
   reactions in presence of palladium complexes, 23–4
transition metals, 119–20, 189
   alkyl derivatives, 7
   ions, 74–88
transition metal-alkyl σ-bonds, 7
trichloroacetic acid, 68, 69
triethylphosphine, 29
trifluoroacetic acid, 12, 17, 21, 24, 64
   oxidation in, 67, 68, 75, 81, 87, 145, 174
trifluoroperacetic acid, 46, 125
trihydrides, 19
2,2,3-trimethylpentane, 53
trimethylphosphine, 31
trimethyl phosphite, 188
tri-n-propylphosphine, 30
triphenylphosphine, 120, 174
triplet carbene, 43
triplet methylene, 44
tungsten, 29
tungstenocene, 26, 184
tungstenocene analogs, 19
tungstenocene derivatives, 19

vanadium, 123, 144, 187
vinyl hydrogen, 31
σ-vinyl Pd complex, 16

zealites, 58
Ziegler–Natta systems, 185–6